全栈自动化测试实战

基于TestNG、HttpClient、Selenium和Appium

卢家涛 / 编著

電子工業出版社
Publishing House of Electronics Industry
北京•BEIJING

内 容 简 介

TestNG 作为 Java 中举足轻重的测试框架，除可以替代 JUnit 进行单元自动化测试外，还可以通过集成 Selenium、Appium 和 HttpClient 等框架做各种类型的自动化测试。本书首先对自动化测试进行了概述，接着对 TestNG 的语法进行了讲解，核心部分是使用 TestNG 进行单元自动化测试、接口自动化测试和界面自动化测试，最后介绍了持续集成、Mock 测试和代码覆盖率等扩展知识。

本书适合测试工程师、自动化测试工程师和测试管理者阅读。

图书在版编目（CIP）数据

全栈自动化测试实战：基于 TestNG、HttpClient、Selenium 和 Appium / 卢家涛编著. —北京：电子工业出版社，2020.3

ISBN 978-7-121-38382-3

Ⅰ. ①全… Ⅱ. ①卢… Ⅲ. ①软件工具－自动检测 Ⅳ. ①TP311.561

中国版本图书馆 CIP 数据核字(2020)第 021845 号

责任编辑：安　娜
印　　刷：北京天宇星印刷厂
装　　订：北京天宇星印刷厂
出版发行：电子工业出版社
　　　　　北京市海淀区万寿路 173 信箱　　邮编：100036
开　　本：787×980　1/16　印张：15.75　字数：318.8 千字
版　　次：2020 年 3 月第 1 版
印　　次：2023 年 1 月第 6 次印刷
定　　价：79.00 元

凡所购买电子工业出版社图书有缺损问题，请向购买书店调换。若书店售缺，请与本社发行部联系，联系及邮购电话：（010）88254888，88258888。

质量投诉请发邮件至 zlts@phei.com.cn，盗版侵权举报请发邮件至 dbqq@phei.com.cn。

本书咨询联系方式：010-51260888-819，faq@phei.com.cn。

前　言

写作背景

2011 年我接触了第一个自动化测试工具 DeviceAnywhere，该工具将真实手机置于云端，通过录制回放执行自动化测试脚本，最终采用图片对比技术实现断言。自此之后，我对自动化测试技术产生了浓厚的兴趣，在之后的 8 年多时间里，我不断学习自动化测试技术，并把它落实到实际项目中。

为什么是 TestNG，而不是 JUnit?

JUnit 主要用于单元测试，而 TestNG 在 JUnit 的基础上做了很多改进，更适合做全栈自动化测试的整体框架。

本书内容

本书的写作顺序不是由浅入深的，而是“自底向上”的，即按照单元自动化测试、接口自动化测试和界面自动化测试的顺序编写的。

本书首先对自动化测试进行了概述（第 1 章）；

接着对 Java 和 TestNG 语法进行了讲解（第 2 章和第 3 章）；

核心部分是使用 TestNG 进行单元自动化测试、接口自动化测试和界面自动化测试（第 4 章至第 10 章）；

最后介绍了持续集成、Mock 测试和代码覆盖率等扩展知识（第 11 章）。

对于代码基础比较薄弱的读者，建议在阅读完前三章内容后，先阅读较为简单的第 5、7、8 章，再阅读第 4、6、9、10、11 章。

适合读者

测试工程师：本书可以从简单的 Web 自动化测试（第 7 章）带你走进自动化测试的大门。

自动化测试工程师：打通各个级别的自动化测试壁垒，帮助读者全面掌握单元自动化测试、接口自动化测试和界面自动化测试。

测试管理者：了解主流自动化测试技术，指导团队开展自动化测试，把握自动化测试的方向和目标。

致谢

感谢老婆的理解和支持，让我能全身心地编写本书。

感谢老大（陈恒骥）给我进入软件测试行业的机会，没有这个机会，我不可能在这个行业走得这么远。

感谢编辑安娜在本书出版过程中耐心的指导。

虽然书中的每个技术点都曾在实际项目中实践过，但由于时间仓促难免有误，敬请读者批评指正。

笔者的 GitHub：https://github.com/lujiatao2/httpinterface/releases。

卢家涛

2019 年 12 月 20 日

目 录

读者服务

微信扫码回复：**38382**

- 获取免费增值资源
- 获取精选书单推荐
- 加入读者交流群，与更多读者互动

第1章 自动化测试概述

1.1 自动化测试的定义和分类

自动化测试通俗地讲就是使用软件 A 自动执行测试用例来测试软件 B。软件 A 既可以是现有的自动化测试工具，也可以是自己编写的测试脚本。软件 B 就是待测试软件。

自动化测试的分类维度如下。

1. 根据测试阶段划分

（1）单元自动化测试

单元自动化测试是指通过自动化手段对软件最小可测单元（模块）进行的测试。一个最小可测单元通常为一个方法或函数。单元自动化测试常用的工具（框架）有 JUnit、TestNG、Jtest、unittest 和 Pytest 等。

（2）集成自动化测试

集成自动化测试是指通过自动化手段把软件的模块组合起来测试。集成测试分为增量测试和非增量测试两种，这两种测试在手工或自动化的集成测试中都可以采用。

① 增量测试。

在增量测试中，如果增量方式是自顶向下的，那么测试的过程需要桩；如果增量方式是自底向上的，那么测试的过程需要驱动。在实际项目中，经常涉及的“打桩”概念就来自于增量测试。

② 非增量测试。

非增量测试不像增量测试那样有严格的增量顺序，因此在非增量测试过程中，可能同时需要桩和驱动，也可能只需要其中一个，或者两者都不需要。

集成自动化测试的主要表现之一为接口自动化测试，用于接口自动化测试的常用工具（框架）有 JMeter、HttpClient、requests、SoapUI 和 Postman 等。

（3）系统自动化测试

系统自动化测试是指通过自动化手段，将软件、硬件、操作人员当成一个整体进行测试。系统自动化测试又可分为功能自动化测试、性能自动化测试、安全（渗透）自动化测试和兼容性自动化测试等。系统自动化测试常用的工具（框架）有很多，比如 Testin 云测平台就提供了兼容性自动化测试功能，它可以测试 App 在各种手机上运行是否正常。

2．根据测试类型划分

（1）功能自动化测试

功能自动化测试是指通过自动化手段检查软件能否达到预期功能的测试。主要表现为 Web 应用、移动应用和桌面应用等软件的界面测试。界面自动化测试的常用工具（框架）有 Selenium、Appium、Robot Framework、UFT/QTP 和 AutoIt 等。

（2）性能自动化测试

性能自动化测试是指通过自动化手段模拟各种正常、峰值和异常负载条件，从而对系统的各项性能指标进行测试。在性能自动化测试中，JMeter 和 LoadRunner 占据了绝对的统治地位。

（3）安全（渗透）自动化测试

安全（渗透）自动化测试可理解为通过自动化手段检查软件是否有安全漏洞。安全（渗透）自动化测试包含前期交互、信息收集、威胁建模、漏洞分析、渗透攻击、后渗透攻击和测试报告 7 个阶段。安全（渗透）自动化测试常用工具（框架）有 Metasploit、Burp Suite、Nessus、NMap、sqlmap、Synopsys Defensics 和 Peach Fuzzer 等。

3．根据静态和动态划分

（1）静态自动化测试

静态自动化测试是指通过自动化手段不实际运行软件的测试，比如代码审查、文档测试等。根据编程语言不同，代码审查自动化测试工具（框架）的差异很大。对 Java 而言，常用的有 FindBugs、PMD 等。

（2）动态自动化测试

动态自动化测试是指通过自动化手段实际运行软件的测试，即通常所说的“软件测试”。动态自动化测试（框架）前面已经介绍很多，这里不再赘述。

从狭义来讲，自动化测试主要指动态的功能自动化测试。

1.2　自动化测试的目的

1.2.1　提高软件质量

既然谈到软件质量，那么就有必要了解一下软件质量的度量标准。

1. 需求覆盖率

需求覆盖率更多的是从产品经理的角度出发，统计产品需求文档中的需求被覆盖了多少，从而计算出需求覆盖率：

需求覆盖率 = 测试覆盖的需求数 / 需求总数 × 100%

2. 代码覆盖率

开发人员可能更关心代码覆盖率。代码覆盖方式有很多种，下面举例说明。

有一个名为 demo 的 Java 方法如下：

```
public void demo(boolean a, boolean b, boolean c, boolean d) {
    if (a && b) {
        System.out.println(true);
    } else if (c || d) {
        System.out.println(false);
    }
}
```

（1）语句覆盖

语句覆盖的原则是覆盖每条语句，针对 demo 方法，一条测试用例即可完成覆盖：

Case 1：a=true，b=true，c=true，d=false

（2）分支（判定）覆盖

语句覆盖并没有考虑 if 语句为假（false）的情况，显然测试并不充分。分支（判定）覆盖可以解决这个问题。采用分支（判定）覆盖重写的测试用例如下：

Case 1：a=true，b=true，c=true，d=false

Case 2：a=true，b=false，c=false，d=false

（3）条件覆盖

分支（判定）覆盖看似比语句覆盖更加完美，但分支（判定）覆盖并没有考虑每个条件的每个取值（即 a、b、c、d 均可以取 true 或 false 两个值）。条件覆盖能覆盖到每个条件的每个取值，采用条件覆盖重写的测试用例如下：

Case 1：a=true，b=false，c=true，d=false

Case 2：a=false，b=true，c=false，d=true

（4）分支（判定）—条件覆盖

如果能同时满足分支（判定）覆盖和条件覆盖就更好了，而分支（判定）—条件覆盖就能做到，采用分支（判定）—条件覆盖重写的测试用例如下：

Case 1：a=true，b=true，c=true，d=true

Case 2：a=false，b=false，c=false，d=false

（5）条件组合覆盖

条件组合覆盖考虑的是覆盖每个分支（判定）中每个条件的每种组合，采用条件覆盖重写的测试用例如下：

Case 1：a=true，b=true，c=true，d=true

Case 2：a=true，b=false，c=true，d=false

Case 3：a=false，b=true，c=false，d=true

Case 4：a=false，b=false，c=false，d=false

下面对这 4 条测试用例进行详细解释：

若 a && b 为 true，那么 a=true，b=true。

若 a && b 为 false，那么 a=true，b=false；或 a=false，b=true；或 a=false，b=false。

若 c && d 为 true，那么 c=true，d=true；或 c=true，d=false；或 c=false，d=true。

若 c && d 为 false，那么 c=false，d=false。

（6）路径覆盖

路径覆盖的原则是覆盖所有路径，针对 demo 方法，共包含 4 条路径：

a && b 为 true，同时 c && d 为 true。

a && b 为 true，同时 c && d 为 false。

a && b 为 false，同时 c && d 为 true。

a && b 为 false，同时 c && d 为 false。

对应测试用例如下：

Case 1：a=true，b=true，c=true，d=false

Case 2：a=true，b=true，c=false，d=false

Case 3：a=true，b=false，c=true，d=false

Case 4：a=true，b=false，c=false，d=false

如何计算代码覆盖率呢？以路径覆盖为例，假如执行了 Case 1，那么路径覆盖率为 25%（即 1 / 4 × 100%）。

关于代码覆盖的更多内容，参见本书第 12 章。

3. 缺陷遗漏率

在软件质量的度量标准中，关于缺陷的不在少数，比如缺陷遗漏率、缺陷修复率、缺陷遗留率和严重缺陷率等。笔者见得最多的是缺陷遗漏率，正所谓“玉瓷之石，金刚试之”，软件质量好不好，用户说了算。当缺陷太多时，说明软件质量仍有很大的提升空间。

缺陷遗漏率 = 线上缺陷数 / 缺陷总数 × 100%

缺陷遗漏率不仅是软件质量的度量标准，也是测试人员考核的硬指标之一。

从软件质量的度量标准中无法看出自动化测试是如何提高质量的，下面举例说明。

假设某软件频繁地进行回归测试，测试人员已经身心疲惫，难以保证每次回归测试的质量，这时如果使用自动化测试代替测试人员进行回归测试，那么回归测试的质量就能得到很好的保证（机器是不知疲倦的），从而减少线上缺陷数，降低缺陷遗漏率。

1.2.2　提高测试效率

可能很多人认为：手工测试执行的用例由自动化测试代替，测试效率的提高显而易见。

但事实上不能如此片面地下结论，原因如下。

（1）自动化测试用例的编写和维护效率远低于手工测试用例

自动化测试用例一天只有几条的产出，而手工测试用例一天可以写几十条，甚至上百条。对于迭代频繁的产品，维护自动化测试用例将是测试人员的“噩梦”。

（2）自动化测试用例执行失败后需要人工分析

自动化测试用例虽然可以做断言，但执行结果仍然需要由测试人员进行人工分析，判断是自动化测试用例本身的问题还是真正的缺陷。

（3）自动化测试无法处理未知的场景

假设测试人员误删除了自动化测试用例在测试环境中创建的一条数据，那么这条自动化测试用例就会执行失败，从而增加人工分析的时间。而手工测试则不同，如果发现创建的数据被删除了，那么再创建一条即可。

从以上几点来看，自动化测试要想提高测试效率，需要做到以下几点。

① 自动化测试建设初期需要投入更多的人力，否则自动化测试无法形成一定的体量，也就谈不上代替手工测试提高测试效率。

② 产品迭代频率不能太快，否则自动化测试用例将难以跟上维护的节奏，几个迭代过后，自动化用例将不可用。

③ 自动化测试用例质量要过关，否则在执行完成后会出现大量的“误报”，即自动化测试用例本身的问题，并非真正的缺陷。

④ 由于自动化测试对执行环境要求非常苛刻，因此不能与手工测试环境混用。

1.3 自动化测试实施三要素

1.3.1 有明确的目标

做事情如果没有目标，那就是盲目的、随意的，做自动化测试也是如此。下面分别举例说明如何制定提高软件质量和测试效率的目标。

1. 制定提高软件质量的目标

以缺陷遗漏率为例。表 1-1 为某项目的线上缺陷分析情况。

表 1-1

缺陷类别	线上缺陷数	测试发现缺陷数	缺陷总数	缺陷遗漏率
前端展示	1	28	29	3.45%
前端逻辑	4	126	130	3.08%
后端逻辑	22	276	298	7.38%
性能缺陷	0	2	2	0.00%
安全缺陷	0	1	1	0.00%
总　　计	27	433	460	5.87%

从表 1-1 可以看出，后端逻辑的线上缺陷数最高，如果能降低后端逻辑的线上缺陷数，则可有效降低缺陷遗漏率（缺陷遗漏率 = 线上缺陷数 / 缺陷总数 × 100%）。

如果采用接口自动化测试来覆盖后端逻辑，实现 50%的接口逻辑全覆盖，那么理论上线上缺陷数将减少一半，即降低为 11 个，最后可算出缺陷遗漏率为 3.56%。此时提高软件质量的目标定为 4%比较合适，毕竟真正做到“逻辑全覆盖”是很难的。

2. 制定提高测试效率的目标

以回归测试效率为例。假设某项目的界面自动化用例约为 1500 个，项目迭代周期为半个月，平均每个迭代周期内需要做两轮回归测试，手工用例执行效率为 100 个/人/天。自动化测试用例采用 5 台执行机并发执行，开发提交测试的当天晚上配置好 5 台执行机开始执行，第二天花 1 人/天即可分析完成。但每次迭代完成后，需要花费约 2 人/天做自动化测试用例的维护，如表 1-2 所示。

表 1-2

测试类型	每迭代所需人力	每月所需人力	每季度所需人力	每年所需人力
手工测试	30	60	180	720
自动化测试	4	8	24	96
节省人力	26	52	156	624

通过表 1-2 可以看出，自动化测试明显节省了人力。下面计算对回归测试效率的提高（以每迭代为例）：

（30 – 4）/ 4 × 100% = 650%

也就是说，当自动化用例数为 1500 个时，回归测试效率是原来的 6.5 倍（即 650%），不过有两点需要说明。

（1）前期投入

这 1500 个自动化用例的编写大约需要一年时间，平均每个工作日投入人力为 2 人，按一年 250 个工作日计算，投入人力为 500 人/天，即一天一个人实现 3 条自动化测试用例。

（2）维护成本

每迭代只需花费 2 人/天做自动化测试用例的维护，可以看出项目界面变更并不频繁。如果项目界面变更很频繁，则势必要增加维护成本。

通过这个例子可以看出，在做自动化测试之前，需要计算项目的回归测试用例数、前期投入和维护成本，才能得出真正提高测试效率的目标。对于界面变更很频繁的项目，做界面自动化测试甚至比手工测试投入的人力更多，对于这种项目，必须果断放弃界面自动化测试。笔者就经历过很多这样的项目。

1.3.2　有足够的资源

当目标明确以后，就初步确认了需要做哪种类型的自动化测试，以及待实现自动化的手工用例数量（工作量）。接下来还要做一件事情，即申请资源。

1. 人力资源

根据测试人员水平的不同，可以做不同程度的自动化，理想情况需要至少一位测试开发工程师和一位自动化测试工程师，其他参与者为普通测试工程师。

测试开发工程师：负责自动化测试整体框架的维护，做必要的扩展开发。

自动化测试工程师：负责底层函数封装，供自动化测试用例组装时直接调用。

普通测试工程师：组装、调试、执行、维护自动化测试用例。

2. 硬件资源

（1）被测系统

被测系统（System Under Test，SUT）需要的服务器数量需要明确，同时服务器的操作系统版本、CPU 大小、内存大小和磁盘大小同样需要确定下来。

（2）自动化测试基础框架

自动化测试基础框架包含 CI/CD 服务器、执行机、SVN/Git 服务器和私服等。这些同样需要明确操作系统版本、CPU 大小、内存大小和磁盘大小。

1.3.3 有合理的计划

计划不是万能的，但没有计划是万万不能的。计划一般分为短期计划和长期计划两种。以采用关键字驱动的界面自动化为例。

1. 短期计划

有目标、任务项、完成时间、责任人，另外还需要明确反馈机制，即反馈频率和反馈内容。反馈频率一般按周反馈比较合适，具体可根据实际情况制定。反馈内容包括公共关键字数、业务关键字数、自动化用例数、自动化率等。对测试开发人员来说，还可以将代码行数加入反馈内容。

2. 长期计划

长期计划比较粗糙，且不确认因素较多，主要展示自动化测试的潜力。

表 1-3 为自动化测试计划示例。

表 1-3

阶　　段	目　　标	任务项	完成时间	责任人	备　　注
第一阶段（短期计划）	1. 搭建自动化测试基础框架 2. 实现P0级别界面自动化（600个用例）	界面自动化测试工具（框架）调研	2019/7/5	张三	
		自动化测试基础框架搭建（CI、执行机等）	2019/7/5	李四	
		SUT搭建	2019/7/5	李四	
		Demo开发及演示	2019/7/12	张三	
		自动化测试用例开发流程及标准制定	2019/7/12	李四	
		公共关键字实现	2019/7/31	张三、李四	
		项目1业务关键字&自动化测试用例实现	2019/12/31	项目1测试	
		项目2业务关键字&自动化测试用例实现	2019/12/31	项目2测试	
		项目3业务关键字&自动化测试用例实现	2019/12/31	项目3测试	
第二阶段（长期计划）	1. 实现P0/P1级别接口自动化 2. 完善自动化测试整体框架	P0级别界面自动化持续维护	待定	待定	
		实现P0/P1级别接口自动化			
		Mock测试接入			
		代码覆盖率接入			

任务项说明如下。

（1）界面自动化测试工具（框架）调研

在确定要做哪种类型的自动化测试后就可以对工具（框架）进行选型，即使之前有成熟的案例，由于新工具（框架）的不断涌现，在实施自动化测试之前也有必要再对比一下。

（2）自动化测试基础框架搭建（CI、执行机等）

资源申请下来后，需要对自动化测试基础框架进行搭建，包括 CI、执行机、SVN/Git 服务器和私服等。CI 一般使用 Jenkins、TeamCity 和 Bamboo 等。Web 测试执行机一般用 Windows 系统的电脑，App 测试执行机还需要用到 macOS 系统的电脑及 Android/iOS 手机。SVN/Git 服务器和私服一般用 Linux 搭建。

（3）SUT 搭建

首先搭建数据库、缓存、MQ 和应用服务器等基础组件，搭建完成后，再将项目部署到新的环境上。这个过程通常需要开发人员配合修改一些必要的配置文件。

（4）Demo 开发及演示

基于自动化测试基础框架及 SUT 编写 Demo，Demo 用于演示技术上的可行性。

（5）自动化测试用例开发流程及标准制定

“不以规矩，不能成方圆”，自动化测试用例的开发必须要有相应的流程和标准来作为指导和约束。流程可参考图 1-1。

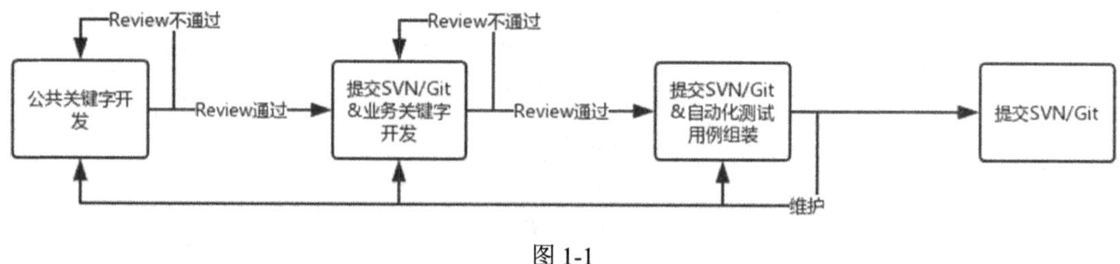

图 1-1

对于标准，这里只介绍几个核心点。

① 用例之间必须解耦。

解耦的目的主要是为了方便维护，以及执行时不相互影响。具体可理解为用例 A 不依赖用例 B，也不影响用例 C。

② 测试用例与测试数据分离。

方便分别维护逻辑（用例）和输入（数据）。

③ 手工测试环境与自动化测试环境分离。

避免手工测试与自动化测试相互影响。

④ 执行策略与执行任务分离。

执行策略包括定时执行、延时执行、周期执行、执行的用例级别等，与具体业务无关。执行策略和执行任务的关系类似于公共关键字和业务关键字的关系。

（6）公共关键字实现

公共关键字与具体业务无关，可在所有项目中复用。

（7）业务关键字与自动化测试用例实现

业务关键字与具体业务有关，仅可在一个或少数项目中复用。自动化测试用例的实现同时依靠公共关键字和业务关键字。

（8）P0 级别界面自动化持续维护

第一阶段实现的自动化测试用例必须持续维护，不然容易“荒废”，因此在第二阶段必然会维护第一阶段实现的自动化测试用例。

（9）实现 P0/P1 级别接口自动化

敏捷大师 Mike Cohn 提出了测试金字塔，如图 1-2 所示，主要思想是多做底层的测试，少做上层的测试，因为底层测试收益高，上层测试收益低。因此界面自动化测试不宜过多投入，精力应该下放到服务层和单元层。所以在第二阶段将接口自动化测试的实现范围扩展到了 P0/P1 级别。

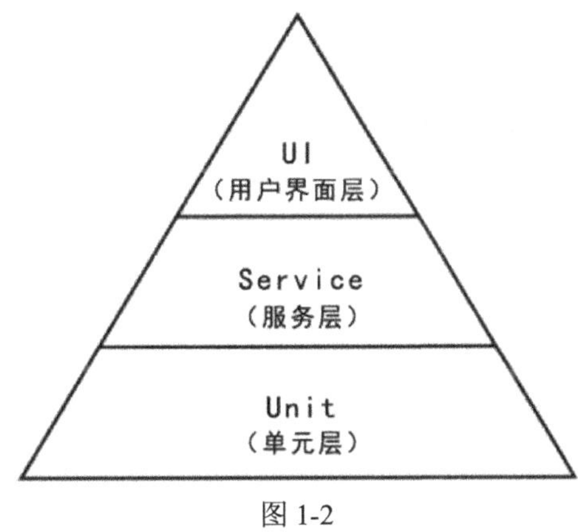

图 1-2

（10）Mock 测试接入

Mock 测试在单元测试中非常常见，另外，在对接第三方服务时往往需要先使用 Mock 测试对内部功能测试完成后，再对接第三方服务联测。

（11）代码覆盖率接入

代码覆盖率是软件质量的度量标准之一，因此非常有必要把代码覆盖率工具（框架）和自动化测试进行整合。

第2章 TestNG 语法

2.1 TestNG 简介

TestNG（Test Next Generation，下一代测试技术）在 JUnit 和 NUnit 基础上新增了许多功能。TestNG 支持单元自动化测试，另外，它还可以集成 Selenium、Appium 和 HttpClient 等框架做各种类型的自动化测试。主要特点如下。

- 强大的注解功能。
- 数据驱动。
- 灵活的测试配置。
- 支持多种并发测试策略。
- 可以和多种工具（插件）协同使用，比如 Eclipse、IntelliJ IDEA 和 Maven 等。

2.2 测试前的准备工作

2.2.1 配置 Maven

在 Java 构建工具中，常用的有 Gradle、Maven 和 Ant。本书统一使用 Maven。Maven 既可以使用 Eclipse 自带的，也可以使用本地计算机上的，为简明步骤，本书使用 Eclipse 自带的 Maven，配置方法如下。

① 在 C:\Users\lujiatao\.m2 目录中新增 settings.xml 文件，输入以下内容。lujiatao 为计算机的用户名，需根据实际情况替换。另外，如果.m2 目录不存在，则自行创建即可：

```
<?xml version="1.0" encoding="UTF-8"?>
<settings xmlns="http://maven.apache.org/SETTINGS/1.0.0"
```

```
    xmlns:xsi="http://www.w3.org/2001/XMLSchema-instance"
    xsi:schemaLocation="http://maven.apache.org/SETTINGS/1.0.0
http://maven.apache.org/xsd/settings-1.0.0.xsd">

    <mirrors>
        <mirror>
            <id>alimaven</id>
            <name>aliyun maven</name>
            <url>http://maven.aliyun.com/nexus/content/groups/public/</url>
            <mirrorOf>central</mirrorOf>
        </mirror>
    </mirrors>

    <profiles>
        <profile>
            <id>jdk-1.8</id>
            <activation>
                <activeByDefault>true</activeByDefault>
                <jdk>1.8</jdk>
            </activation>
            <properties>
                <maven.compiler.source>1.8</maven.compiler.source>
                <maven.compiler.target>1.8</maven.compiler.target>

    <maven.compiler.compilerVersion>1.8</maven.compiler.compilerVersion>
            </properties>
        </profile>
    </profiles>

</settings>
```

下面对上述内容进行说明。

- <mirror>标签用于配置镜像地址，如果不配置，则默认会到 Maven 中央仓库下载依赖文件。由于 Maven 中央仓库服务器在国外，下载速度较慢，因此这里将镜像地址配置在了国内的阿里云上，这样下载速度会快很多。
- <profile>标签主要用于动态配置多套环境，这里用于配置 JDK 版本。如果不配置，则 Maven 默认使用 JDK 5。

② 在 Eclipse 中单击“Window → Preferences → Maven → User Settings→ Browse”，选择“C:\Users\lujiatao\.m2\settings.xml”文件，如图 2-1 所示。

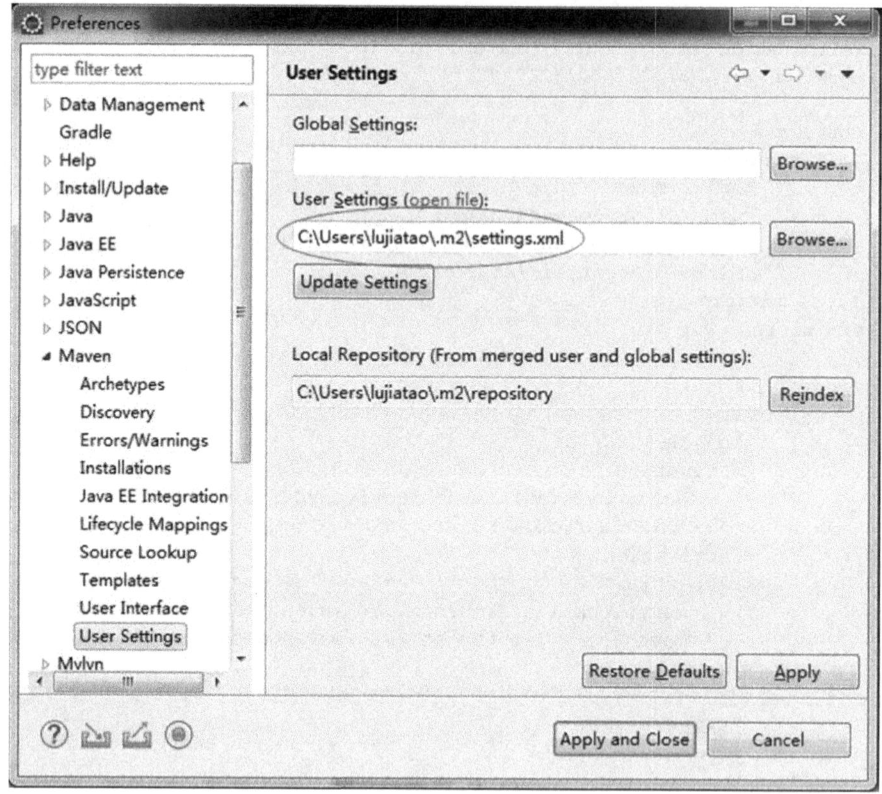

图 2-1

③最后依次单击“Update Settings”“Apply and Close”按钮完成配置。

2.2.2 创建工程

单击“File → New → Project”选项，打开如图 2-2 所示的对话框。

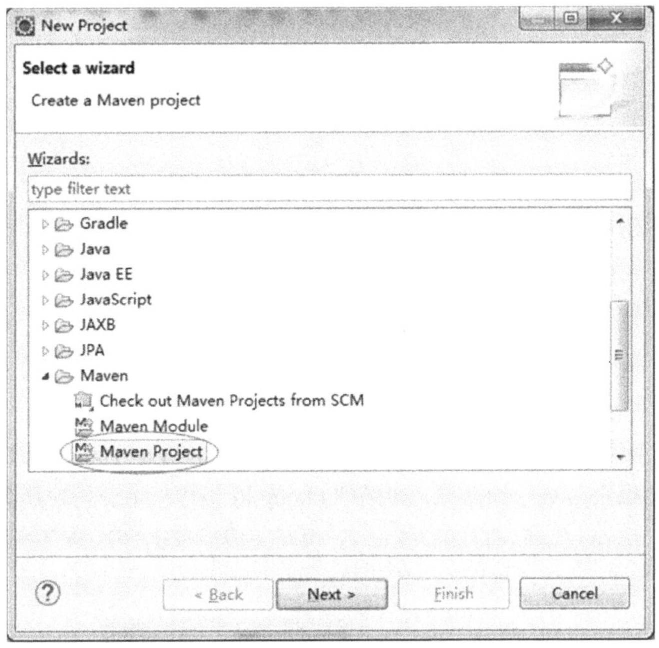

图 2-2

选择“Maven → Maven Project”选项，单击“Next”按钮进入下一步，如图 2-3 所示。

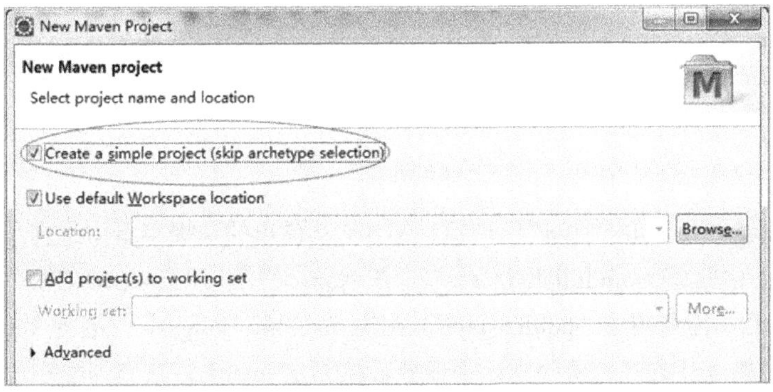

图 2-3

勾选“Create a simple project(skip archetype selection)”，该选项表示只需创建一个简单的 Maven 工程，不需要任何模板。单击“Next”按钮进入下一步，如图 2-4 所示。

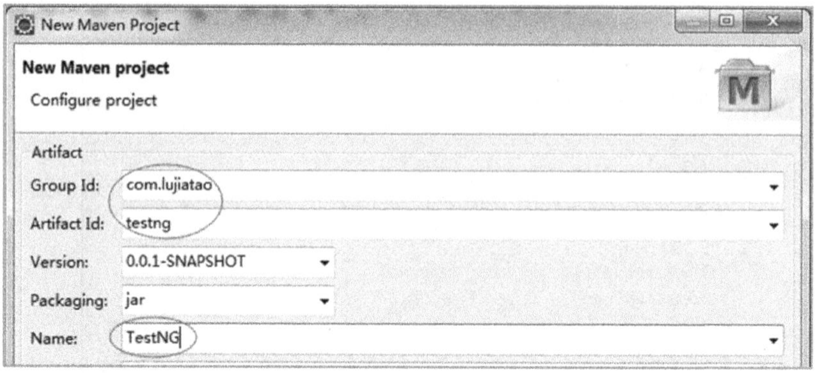

图 2-4

Group Id 代表组织的唯一标识符，采用反向域名（和 Java 的 Package 命名类似）的形式。Artifact Id 代表该组织内的唯一标识符。如果把 Group Id 当作一个公司，那么 Artifact Id 就是该公司中的一个项目。Name 代表项目的完整命名。填写完成后单击“Finish”按钮，一个名为“testng”的工程就出现在了 Eclipse 中，如图 2-5 所示。

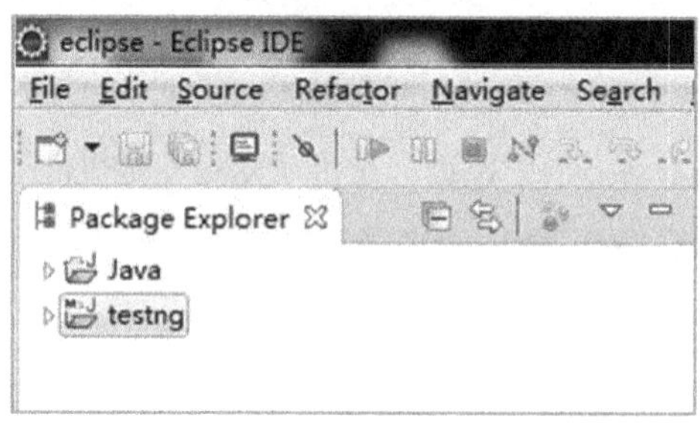

图 2-5

最后在“testng → src/test/java”上用鼠标右击，从弹出的快捷菜单中选择“New → Class”选项。在 Package 栏填写“com.lujiatao.testng”，在 Name 栏填写“FirstClassTest”，单击“Finish”按钮。注意，此时不需要勾选“public static void main(String[] args)”，因为 TestNG 的运行方式和普通 Java 项目的运行方式有所不同，不需要显式指定 main()方法作为程序运行的入口。

2.2.3 测试执行

1. 配置 TestNG 依赖

想要用 TestNG 执行测试用例，首先要有 TestNG。下面使用 Maven 自动下载 TestNG，具体操作是在 pom.xml 文件的<name>标签后输入以下粗体部分内容：

```
<project xmlns="http://maven.apache.org/POM/4.0.0"
    xmlns:xsi="http://www.w3.org/2001/XMLSchema-instance"
    xsi:schemaLocation="http://maven.apache.org/POM/4.0.0
```

```
http://maven.apache.org/xsd/maven-4.0.0.xsd">
    <modelVersion>4.0.0</modelVersion>
    <groupId>com.lujiatao</groupId>
    <artifactId>testng</artifactId>
    <version>0.0.1-SNAPSHOT</version>
    <name>TestNG</name>

    <dependencies>
        <dependency>
            <groupId>org.testng</groupId>
            <artifactId>testng</artifactId>
            <version>6.14.3</version>
            <scope>test</scope>
        </dependency>
    </dependencies>

</project>
```

保存 pom.xml 文件，这时 Maven 会自动下载 TestNG 及其依赖的其他 jar 包。

2. 安装 TestNG 插件

在 Eclipse 中单击“Help →Install New Software”，在 Work with 中输入“http://beust.com/eclipse”并按回车键，稍等片刻后下方会出现“TestNG”，勾选“TestNG”，如图 2-6 所示，单击“Next”按钮进入下一步，如图 2-7 所示。

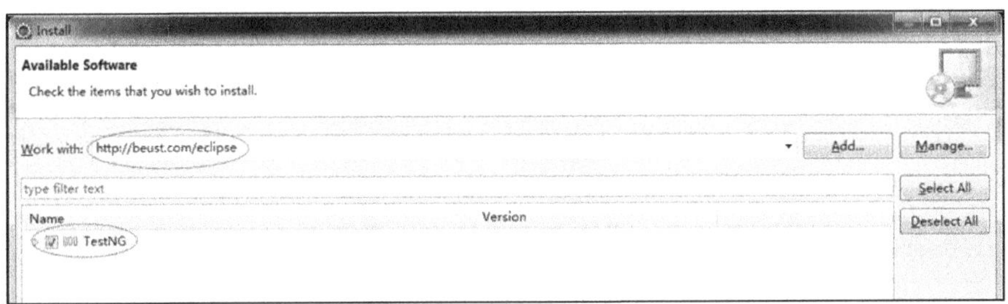

图 2-6

图 2-7

这里显示了要安装的插件名称、版本和 ID。单击“Next”按钮进入下一步，这时会显示 Apache 的许可信息，勾选“I accept the terms of the license agreement”，表示同意许可信息，单击“Finish”按钮。插件开始自动下载并安装，安装完成后会提示重启 Eclipse，重启即可。

3. 编写和执行测试用例

在 FirstClassTest 中输入以下粗体部分内容：

```
package com.lujiatao.testng;

import org.testng.annotations.Test;

public class FirstClassTest {

    @Test
    public void testCase1() {
        System.out.println("Hello TestNG!");
    }

}
```

保存代码，在 FirstClassTest.java 上用鼠标右击，从弹出的快捷菜单中选择“Run As → TestNG Test” 选项，此时 Eclipse 的控制台输出如下：

```
[RemoteTestNG] detected TestNG version 6.14.3
Hello TestNG!
PASSED: testCase1

===============================================
    Default test
    Tests run: 1, Failures: 0, Skips: 0
===============================================


===============================================
Default suite
Total tests run: 1, Failures: 0, Skips: 0
===============================================
```

下面对运行结果进行说明。

① 第一行显示了 TestNG 的版本号。

② 使用 System.out.println()方法打印“Hello TestNG!”。

③ 测试用例 testCase1 的测试结果为通过（PASSED）。在 TestNG 中，一个用@Test 注解修饰的方法代表一个测试用例。

④ 一个 Test 代表一系列 Class 的合集，此处默认为 Default test。共运行 1 条测试用例，其中失败 0 条，跳过 0 条。

⑤ 一个 Suite 代表一系列 Test 的合集，此处默认为 Default suite。共运行 1 条测试用例，其中失败 0 条，跳过 0 条。

4. 查看测试报告

运行 TestNG 后，工程中会多出一个 test-output 目录。展开 test-output 目录，在 emailable-report.html 上用鼠标右击，从弹出的快捷菜单中选择“Open With → Web Browser”选项，此时可以看到 HTML 格式的测试报告，如图 2-8 所示。

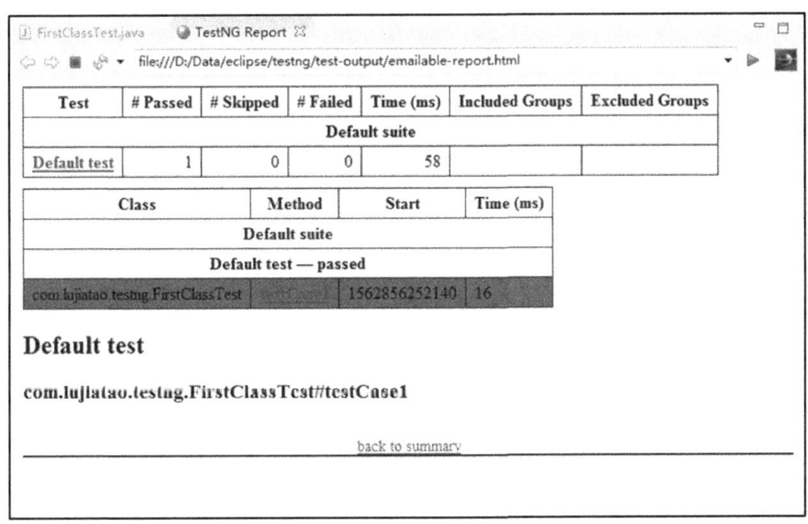

图 2-8

除默认测试报告外，TestNG 还支持与第三方测试报告的集成，常见的包括 ReportNG、Extent Reporting 和 Allure 等。当然，TestNG 也支持自定义测试报告。

2.3 TestNG 注解

TestNG 和其他很多 Java 框架（比如 JUnit、Spring 等）一样，使用了大量的注解。被不同注解修饰的类、方法具有不同的含义，本节对 TestNG 注解进行详细介绍，并按照使用场景把注解分成 4 类。

（1）前置条件和后置条件

把注解作为前置条件（或初始化操作）和后置条件（或清理操作）使用。

（2）数据驱动

TestNG 的特点之一是数据驱动，即测试用例和测试数据分离，以便维护和管理。

（3）测试用例

该分类只有一个@Test 注解。@Test 注解的作用是对测试用例进行控制，该注解中的方法有很多，后面会对常用方法进行介绍。

（4）监听器

该分类只有一个@Listeners 注解。监听器的作用是监控测试过程，如果采用默认监听器，则不需要任何配置；如果使用自定义监听器，则需要使用@Listeners 注解或 testng.xml 文件进行配置。由于篇幅所限，本章不对自定义监听器进行介绍，有兴趣的读者可以自行查阅相关资料。

2.3.1　前置条件和后置条件

先来看看各注解的含义。

@BeforeSuite：在该 Suite 的所有 Test 都未运行之前运行。

@AfterSuite：在该 Suite 的所有 Test 都运行之后运行。

一个 Suite 对应一个顶级模块，比如一个软件项目分为 4 个模块，那么每个模块就是一个 Suite。一般结合 testng.xml 文件中的<suite>或<suite-files>标签进行使用。

@BeforeTest：在该 Test 的所有 Class 都未运行之前运行。

@AfterTest：在该 Test 的所有 Class 都运行之后运行。

一个 Test 对应一个子模块，一般结合 testng.xml 文件中的<test>标签进行使用。

@BeforeClass：在该 Class 的所有@Test 方法都未运行之前运行。

@AfterClass：在该 Class 的所有@Test 方法都运行之后运行。

一个 Class 对应一个 Java 类，在该 Java 类中，用@BeforeClass（或@AfterClass）修饰的方法会在该 Class 的所有@Test 方法都运行之前（或之后）运行。

@BeforeMethod：在该 Class 的每个@Test 方法运行之前运行。

@AfterMethod：在该 Class 的每个@Test 方法运行之后运行。

一个 Method 对应一个 Java 方法，在 Java 类中用@BeforeMethod（或@AfterMethod）修饰的方法会在该 Class 的每个@Test 方法运行之前（或之后）运行。

@BeforeGroups：在该 Class 第一个分组的@Test 方法运行之前运行。

@AfterGroups：在该 Class 最后一个分组的@Test 方法运行之后运行。

Group 的控制粒度介于 Class 和 Method 之间，一个 Class 可以包含多个 Group，一个 Group 可以包含多个 Method。

只看文字是很生硬的，下面通过例子来说明以上注解。删除 FirstClassTest 中的内容，输入以下代码：

```
package com.lujiatao.testng;
```

```
import org.testng.annotations.AfterClass;
import org.testng.annotations.AfterGroups;
import org.testng.annotations.AfterMethod;
import org.testng.annotations.AfterSuite;
import org.testng.annotations.AfterTest;
import org.testng.annotations.BeforeClass;
import org.testng.annotations.BeforeGroups;
import org.testng.annotations.BeforeMethod;
import org.testng.annotations.BeforeSuite;
import org.testng.annotations.BeforeTest;
import org.testng.annotations.Test;

public class FirstClassTest {

    @BeforeSuite
    public void beforeSuite() {
        System.out.println("beforeSuite");
    }

    @AfterSuite
    public void afterSuite() {
        System.out.println("afterSuite");
    }

    @BeforeTest
    public void beforeTest() {
        System.out.println("++beforeTest");
    }

    @AfterTest
    public void afterTest() {
        System.out.println("++afterTest");
    }

    @BeforeClass
    public void beforeClass() {
        System.out.println("++--beforeClass");
    }

    @AfterClass
    public void afterClass() {
        System.out.println("++--afterClass");
    }

    @BeforeGroups(groups = { "g1" })
    public void beforeGroups1() {
        System.out.println("++--++beforeGroups1");
    }

    @AfterGroups(groups = { "g1" })
```

```
    public void afterGroups1() {
        System.out.println("++--++afterGroups1");
    }

    @BeforeGroups(groups = { "g2" })
    public void beforeGroups2() {
        System.out.println("++--++beforeGroups2");
    }

    @AfterGroups(groups = { "g2" })
    public void afterGroups2() {
        System.out.println("++--++afterGroups2");
    }

    @BeforeMethod
    public void beforeMethod() {
        System.out.println("++--++--beforeMethod");
    }

    @AfterMethod
    public void afterMethod() {
        System.out.println("++--++--afterMethod");
    }

    @Test(groups = { "g1" })
    public void testCase1() {
        System.out.println("++--++--++testCase1");
    }

    @Test(groups = { "g2" })
    public void testCase2() {
        System.out.println("++--++--++testCase2");
    }

    @Test(groups = { "g2" })
    public void testCase3() {
        System.out.println("++--++--++testCase3");
    }

    @Test(groups = { "g1", "g2" })
    public void testCase4() {
        System.out.println("++--++--++testCase4");
    }

}
```

保存代码，在“FirstClassTest.java”上用鼠标右击，从弹出的快捷菜单中选择“Run As → TestNG Test”选项，此时 Eclipse 的控制台输出如下：

```
[RemoteTestNG] detected TestNG version 6.14.3
beforeSuite
```

```
++beforeTest
++--beforeClass
++--++beforeGroups1
++--++--beforeMethod
++--++--++testCase1
++--++--afterMethod
++--++beforeGroups2
++--++--beforeMethod
++--++--++testCase2
++--++--afterMethod
++--++--beforeMethod
++--++--++testCase3
++--++--afterMethod
++--++--beforeMethod
++--++--++testCase4
++--++--afterMethod
++--++afterGroups2
++--++afterGroups1
++--afterClass
++afterTest
PASSED: testCase1
PASSED: testCase2
PASSED: testCase3
PASSED: testCase4

===============================================
   Default test
   Tests run: 4, Failures: 0, Skips: 0
===============================================

afterSuite

===============================================
Default suite
Total tests run: 4, Failures: 0, Skips: 0
===============================================
```

下面对运行结果进行说明。

① 可以看出@BeforeSuite、@AfterSuite、@BeforeTest、@AfterTest、@BeforeClass 和@AfterClass 控制测试执行的粒度是不同的，即 Suite > Test > Class。

② 一个测试用例(@Test 修饰的 Java 方法)可以属于多个分组，比如上面示例的 testCase4。当多个分组都设置了对应的@BeforeGroups 和@AfterGroups 注解时，执行顺序是 Before Group1→Before Group2→After Group2→After Group1。

③ @BeforeMethod 和@AfterMethod 共执行了 4 次，因为有 4 个测试用例。

2.3.2　数据驱动

TestNG 做数据驱动时使用了@DataProvider 和@Parameters 注解，后者需要和 testng.xml 文件配合。举一个登录的场景，每种不同的输入都对应了不同的提示。

删除 FirstClassTest 中的内容，输入以下代码：

```
package com.lujiatao.testng;

import org.testng.annotations.DataProvider;
import org.testng.annotations.Test;

public class FirstClassTest {

    @Test(dataProvider = "data")
    public void testCase1(String username, String password, String prompt) {
        System.out.println("如果输入:" + username + "、" + password + ",提示" + prompt);
    }

    @DataProvider(name = "data")
    public Object[][] dataProvider1() {
        return new Object[][] { new Object[] { "空账号", "正确密码", "账号不能为空!
" }, new Object[] { "正确账号", "空密码", "密码不能为空! " },
                new Object[] { "正确账号", "正确密码", "登录成功! " } };
    }

}
```

保存代码，在 FirstClassTest.java 上用鼠标右击，从弹出的快捷菜单中选择“Run As → TestNG Test” 选项，此时 Eclipse 的控制台输出如下：

```
[RemoteTestNG] detected TestNG version 6.14.3
如果输入：空账号、正确密码，提示账号不能为空!
如果输入：正确账号、空密码，提示密码不能为空!
如果输入：正确账号、正确密码，提示登录成功!
PASSED: testCase1("空账号", "正确密码", "账号不能为空! ")
PASSED: testCase1("正确账号", "空密码", "密码不能为空! ")
PASSED: testCase1("正确账号", "正确密码", "登录成功! ")

===============================================
    Default test
    Tests run: 3, Failures: 0, Skips: 0
===============================================


===============================================
Default suite
Total tests run: 3, Failures: 0, Skips: 0
===============================================
```

被@DataProvider 修饰的 Java 方法称为数据提供者，该方法返回一个对象二维数组。如果一个测试用例需要该数据，那么就通过@Test 注解的 dataProvider 方法传入数据提供者的名称。

2.3.3 测试用例

@Test 注解的方法很多，前面已经介绍过 groups 和 dataProvider 了，下面再介绍几种常用的。

删除 FirstClassTest 中的内容，输入以下代码：

```
package com.lujiatao.testng;

import org.testng.annotations.Test;

public class FirstClassTest {

    @Test(description = "测试用例 1")
    public void testCase1() {
        System.out.println("testCase1");
    }

    @Test(priority = 2)
    public void testCase2() {
        System.out.println("testCase2");
    }

    @Test(priority = 1)
    public void testCase3() {
        System.out.println("testCase3");
    }

    @Test
    public void testCase4() {
        System.out.println("testCase4");
        throw new RuntimeException("testCase4 运行异常! ");
    }

    @Test(groups = { "myGroup" })
    public void testCase5() {
        System.out.println("testCase5");
        throw new RuntimeException("testCas5 运行异常! ");
    }

    @Test(enabled = false)
    public void testCase6() {
        System.out.println("testCase6");
    }

    @Test(dependsOnMethods = { "testCase4" }, dependsOnGroups = { "myGroup" },
```

```
alwaysRun = true)
    public void testCase7() {
        System.out.println("testCase7");
    }

}
```

保存代码，在“FirstClassTest.java”上用鼠标右击，从弹出的快捷菜单中选择“Run As → TestNG Test” 选项，此时 Eclipse 的控制台输出如下：

```
[RemoteTestNG] detected TestNG version 6.14.3
testCase1
testCase4
testCase5
testCase7
testCase3
testCase2
PASSED: testCase1
        测试用例 1
PASSED: testCase7
PASSED: testCase3
PASSED: testCase2
FAILED: testCase4
java.lang.RuntimeException: testCase4 运行异常!
    at com.lujiatao.testng.FirstClassTest.testCase4(FirstClassTest.java:25)
    at sun.reflect.NativeMethodAccessorImpl.invoke0(Native Method)
    at sun.reflect.NativeMethodAccessorImpl.invoke(Unknown Source)
    at sun.reflect.DelegatingMethodAccessorImpl.invoke(Unknown Source)
    at java.lang.reflect.Method.invoke(Unknown Source)
    at org.testng.internal.MethodInvocationHelper.
        invokeMethod(MethodInvocationHelper.java:124)
    at org.testng.internal.Invoker.invokeMethod(Invoker.java:583)
    at org.testng.internal.Invoker.invokeTestMethod(Invoker.java:719)
    at org.testng.internal.Invoker.invokeTestMethods(Invoker.java:989)
    at org.testng.internal.TestMethodWorker.
        invokeTestMethods(TestMethodWorker.java:125)
    at org.testng.internal.TestMethodWorker.run(TestMethodWorker.java:109)
    at org.testng.TestRunner.privateRun(TestRunner.java:648)
    at org.testng.TestRunner.run(TestRunner.java:505)
    at org.testng.SuiteRunner.runTest(SuiteRunner.java:455)
    at org.testng.SuiteRunner.runSequentially(SuiteRunner.java:450)
    at org.testng.SuiteRunner.privateRun(SuiteRunner.java:415)
    at org.testng.SuiteRunner.run(SuiteRunner.java:364)
    at org.testng.SuiteRunnerWorker.runSuite(SuiteRunnerWorker.java:52)
    at org.testng.SuiteRunnerWorker.run(SuiteRunnerWorker.java:84)
    at org.testng.TestNG.runSuitesSequentially(TestNG.java:1208)
    at org.testng.TestNG.runSuitesLocally(TestNG.java:1137)
    at org.testng.TestNG.runSuites(TestNG.java:1049)
    at org.testng.TestNG.run(TestNG.java:1017)
    at org.testng.remote.AbstractRemoteTestNG.
        run(AbstractRemoteTestNG.java:115)
```

```
    at org.testng.remote.RemoteTestNG.initAndRun(RemoteTestNG.java:251)
    at org.testng.remote.RemoteTestNG.main(RemoteTestNG.java:77)

FAILED: testCase5
java.lang.RuntimeException: testCas5 运行异常!
    at com.lujiatao.testng.FirstClassTest.testCase5(FirstClassTest.java:31)
    at sun.reflect.NativeMethodAccessorImpl.invoke0(Native Method)
    at sun.reflect.NativeMethodAccessorImpl.invoke(Unknown Source)
    at sun.reflect.DelegatingMethodAccessorImpl.invoke(Unknown Source)
    at java.lang.reflect.Method.invoke(Unknown Source)
    at org.testng.internal.MethodInvocationHelper.
      invokeMethod(MethodInvocationHelper.java:124)
    at org.testng.internal.Invoker.invokeMethod(Invoker.java:583)
    at org.testng.internal.Invoker.invokeTestMethod(Invoker.java:719)
    at org.testng.internal.Invoker.invokeTestMethods(Invoker.java:989)
    at org.testng.internal.TestMethodWorker.
      invokeTestMethods(TestMethodWorker.java:125)
    at org.testng.internal.TestMethodWorker.run(TestMethodWorker.java:109)
    at org.testng.TestRunner.privateRun(TestRunner.java:648)
    at org.testng.TestRunner.run(TestRunner.java:505)
    at org.testng.SuiteRunner.runTest(SuiteRunner.java:455)
    at org.testng.SuiteRunner.runSequentially(SuiteRunner.java:450)
    at org.testng.SuiteRunner.privateRun(SuiteRunner.java:415)
    at org.testng.SuiteRunner.run(SuiteRunner.java:364)
    at org.testng.SuiteRunnerWorker.runSuite(SuiteRunnerWorker.java:52)
    at org.testng.SuiteRunnerWorker.run(SuiteRunnerWorker.java:84)
    at org.testng.TestNG.runSuitesSequentially(TestNG.java:1208)
    at org.testng.TestNG.runSuitesLocally(TestNG.java:1137)
    at org.testng.TestNG.runSuites(TestNG.java:1049)
    at org.testng.TestNG.run(TestNG.java:1017)
    at org.testng.remote.AbstractRemoteTestNG.
      run(AbstractRemoteTestNG.java:115)
    at org.testng.remote.RemoteTestNG.initAndRun(RemoteTestNG.java:251)
    at org.testng.remote.RemoteTestNG.main(RemoteTestNG.java:77)


===============================================
  Default test
  Tests run: 6, Failures: 2, Skips: 0
===============================================


===============================================
Default suite
Total tests run: 6, Failures: 2, Skips: 0
===============================================
```

下面对运行结果进行说明。

① description 代表测试用例描述，控制台会打印输出该描述。

② priority 代表优先级，数字越小，优先级越高，默认值为 0。testCase2 的 priority 值为 2，会最后一个执行；testCase3 的 priority 值为 1，会倒数第二个执行。如果级别一样，则执行顺序默认按方法名排序。

③ enabled 的默认值为 true，代表不启用。当 enabled 的值为 false 时，表示禁用，因此 testCase6 并未执行。

④ testCase4 和 testCase5 都抛出了运行时异常，因此执行失败。

⑤ dependsOnMethods 代表依赖一个或多个方法，dependsOnGroups 代表依赖一个或多个分组。一旦被依赖的测试用例执行失败，则 TestNG 将跳过该测试用例。但没有跳过 testCase7，原因是 testCase7 加了 alwaysRun 方法，并将值设为 true，代表始终执行，在默认情况下，该值为 false。建议尽量不要使用 dependsOnMethods 和 dependsOnGroups，因为违背了测试用例需要解耦的原则。

2.4 testng.xml

testng.xml 文件的作用是控制测试执行的过程，该文件可以使用 Eclipse 的 TestNG 插件自动生成。在工程（testng）上用鼠标右击，从弹出的快捷菜单中选择“TestNG → Convert to TestNG”选项，此时会弹出“Refactoring”对话框，直接单击“Finish”按钮，生成“testng.xml”文件，文件内容如下：

```
<?xml version="1.0" encoding="UTF-8"?>
<!DOCTYPE suite SYSTEM "http://testng.org/testng-1.0.dtd">
<suite name="Suite">
    <test thread-count="5" name="Test">
        <classes>
            <class name="com.lujiatao.testng.FirstClassTest" />
        </classes>
    </test> <!-- Test -->
</suite> <!-- Suite -->
```

这是一个典型的 XML 文件，第一行包含了 XML 的声明。<suite>代表一个 Suite；<test>代表一个 Test；<classes>代表一组 Class，可以包含多个 Class，这里默认添加了 FirstClassTest。<test>标签里有一个 thread-count 属性表示并行线程数，该属性需要与 parallel 属性配合使用。

2.4.1 <package>

如果一个 Class 需要一个<class>标签，那么 10 个 Class 就需要 10 个<class>标签。如果这些 Class 都在一个 Package 中，则可以使用<package>标签进行 Package 设置，以简化配置过程。

删除 FirstClassTest 中的内容，输入以下代码：

```
package com.lujiatao.testng;

import org.testng.annotations.Test;

public class FirstClassTest {

    @Test
    public void testCase1() {
        System.out.println("FirstClassTest 的 testCase1");
    }

}
```

新增 SecondClassTest，输入以下内容：

```
package com.lujiatao.testng;

import org.testng.annotations.Test;

public class SecondClassTest {

    @Test
    public void testCase1() {
        System.out.println("SecondClassTest 的 testCase1");
    }

}
```

修改 testng.xml 文件，删除<classes>标签及其内部的配置，并以包设置代替，见以下粗体部分内容：

```
<?xml version="1.0" encoding="UTF-8"?>
<!DOCTYPE suite SYSTEM "http://testng.org/testng-1.0.dtd">
<suite name="Suite">
    <test thread-count="5" name="Test">
        <packages>
            <package name="com.lujiatao.testng" />
        </packages>
    </test> <!-- Test -->
</suite> <!-- Suite -->
```

保存所做的修改，在“testng.xml”上用鼠标右击，从弹出的快捷菜单中选择“Run As → TestNG Suite”选项，此时 Eclipse 的控制台输出如下：

```
[RemoteTestNG] detected TestNG version 6.14.3
FirstClassTest 的 testCase1
SecondClassTest 的 testCase1

===============================================
Suite
Total tests run: 2, Failures: 0, Skips: 0
===============================================
```

从输出结果可以看到两个 Class 都执行了，说明包设置生效。

2.4.2　<include>和<exclude>

<include>和<exclude>的作用是对方法进行设置，<include>代表包含，<exclude>代表排除，写法上它们都支持正则表达式。

删除 FirstClassTest 中的内容，输入以下代码：

```
package com.lujiatao.testng;

import org.testng.annotations.Test;

public class FirstClassTest {

    @Test
    public void testCase1() {
        System.out.println("testCase1");
    }

    @Test
    public void testCase2() {
        System.out.println("testCase2");
    }

    @Test
    public void testCase3() {
        System.out.println("testCase3");
    }

    @Test
    public void newTestCase1() {
        System.out.println("newTestCase1");
    }

}
```

修改 testng.xml 文件，删除<packages>标签及其内部的配置，并以方法设置代替，见以下粗体部分内容：

```
<?xml version="1.0" encoding="UTF-8"?>
<!DOCTYPE suite SYSTEM "http://testng.org/testng-1.0.dtd">
<suite name="Suite">
    <test thread-count="5" name="Test">
        <classes>
            <class name="com.lujiatao.testng.FirstClassTest">
                <methods>
                    <include name="testCase." />
                    <exclude name="testCase3" />
                </methods>
```

```
            </class>
        </classes>
    </test> <!-- Test -->
</suite> <!-- Suite -->
```

保存所做的修改，在“testng.xml”上用鼠标右击，从弹出的快捷菜单中选择“Run As → TestNG Suite”选项，此时 Eclipse 的控制台输出如下：

```
[RemoteTestNG] detected TestNG version 6.14.3
testCase1
testCase2

===============================================
Suite
Total tests run: 2, Failures: 0, Skips: 0
===============================================
```

FirstClass 里共有 testCase1、testCase2、testCase3 和 newTestCase1 四个测试用例，在<include>标签中，name 属性的值为“testCase.”，其中“.”代表任意一个字符，因此 testCase1、testCase2 和 testCase3 都满足要求。但是在<exclude>标签中，name 属性的值为“testCase3”，也就是排除 testCase3。所以最终只有 testCase1 和 testCase2 运行了。

<include>和<exclude>标签除了用在 Test 中，还可以用在其他地方，比如分组运行。删除 FirstClassTest 中的内容，输入以下代码：

```
package com.lujiatao.testng;

import org.testng.annotations.Test;

public class FirstClassTest {

    @Test(groups = { "myGroup" })
    public void testCase1() {
        System.out.println("testCase1");
    }

    @Test
    public void testCase2() {
        System.out.println("testCase2");
    }

    @Test(groups = { "myGroup" })
    public void testCase3() {
        System.out.println("testCase3");
    }

}
```

修改 testng.xml 文件，新增<groups>标签及相关配置，同时修改<classes>标签内的配置，见以下粗体部分内容：

```
<?xml version="1.0" encoding="UTF-8"?>
<!DOCTYPE suite SYSTEM "http://testng.org/testng-1.0.dtd">
<suite name="Suite">
    <groups>
        <run>
            <include name="myGroup" />
        </run>
    </groups>
    <test thread-count="5" name="Test">
        <classes>
            <class name="com.lujiatao.testng.FirstClassTest" />
        </classes>
    </test> <!-- Test -->
</suite> <!-- Suite -->
```

保存所做的修改，在“testng.xml”上用鼠标右击，从弹出的快捷菜单中选择“Run As → TestNG Suite”选项，此时 Eclipse 的控制台输出如下：

```
[RemoteTestNG] detected TestNG version 6.14.3
testCase1
testCase3

===============================================
Suite
Total tests run: 2, Failures: 0, Skips: 0
===============================================
```

可以看到只执行了 myGroup 分组的两条用例，符合预期。

2.4.3 <parameter>标签

<parameter>标签和@Parameters 注解配合使用，可对测试用例传递参数，达到数据分离的效果。删除 FirstClassTest 中的内容，输入以下代码：

```
package com.lujiatao.testng;

import org.testng.annotations.Parameters;
import org.testng.annotations.Test;

public class FirstClassTest {

    @Parameters({ "myParam" })
    @Test
    public void testCase1(String param) {
        System.out.println("myParam 的值为" + param);
    }

}
```

修改 testng.xml 文件，删除<groups>标签及相关配置，同时新增<parameter>标签，详见以下粗体部分内容：

```
<?xml version="1.0" encoding="UTF-8"?>
<!DOCTYPE suite SYSTEM "http://testng.org/testng-1.0.dtd">
<suite name="Suite">
    <parameter name="myParam" value="myParamValue" />
    <test thread-count="5" name="Test">
        <classes>
            <class name="com.lujiatao.testng.FirstClassTest" />
        </classes>
    </test> <!-- Test -->
</suite> <!-- Suite -->
```

保存所做的修改，在“testng.xml”上用鼠标右击，从弹出的快捷菜单中选择“Run As → TestNG Suite”选项，此时 Eclipse 的控制台输出如下：

```
[RemoteTestNG] detected TestNG version 6.14.3
myParam 的值为 myParamValue

===============================================
Suite
Total tests run: 1, Failures: 0, Skips: 0
===============================================
```

可以看到 myParam 的值通过<parameter>标签和@Parameters 注解配合传递给了 testCase1。

第 3 章

单元自动化测试

3.1 编写待测程序

在编写待测程序之前先创建一个新的 Maven 项目，关键信息填写如下。

Group Id：com.lujiatao

Artifact Id：unittest

Name：Unit Test

可根据实际情况填写，不需要完全一致。

在实际项目中，src/main/java 目录用于存放待测程序（程序代码），src/test/java 目录用于存放测试用例（测试代码）。

在 src/main/java 中创建名为 com.lujiatao.unittest 的 Package，以及名为 CalculatorForPpi 的 Class，在 CalculatorForPpi 中输入以下粗体部分内容：

```
package com.lujiatao.unittest;

public class CalculatorForPpi {

    public static long calculate(int width, int height, double size) {
        long result;
        if (width > 0 && height > 0 && size > 0) {
            result = Math.round(Math.pow((Math.pow(width, 2) + Math.pow(height,
                                2)) / Math.pow(size, 2), 0.5));
        } else {
            result = -1;
        }
        return result;
```

```
        }
    }
```

上面编写了一个像素密度（Pixels Per Inch，PPI）计算器，该程序很简单，仅包含一个类和一个方法。实现的功能是，在用户输入屏幕宽、高和尺寸后，计算屏幕的像素密度。

3.2 手工测试用例设计

3.2.1 分析待测程序

1. 分析入参

width 代表屏幕宽，height 代表屏幕高，size 代表屏幕尺寸。其中，屏幕宽和屏幕高入参类型为 int，屏幕尺寸入参类型为 double。

2. 分析返回值

返回屏幕像素密度，返回值类型为 long。

3. 分析程序逻辑

如果屏幕宽、屏幕高和屏幕尺寸任意一个小于或等于 0，则返回值为-1；如果都大于 0，则计算屏幕像素密度。屏幕像素密码的计算公式如下：

$$\text{PPI} = \frac{\sqrt{宽^2 + 高^2}}{尺寸}$$

3.2.2 测试用例设计

通过以上分析，可以编写 1 条计算像素密度的测试用例，编写 6 条不计算像素密度的测试用例，如下所示：

Case 1：width=750，height=1334，size=4.7——计算像素密度

Case 2：width=−1，height=1334，size=4.7——width<0，不计算像素密度

Case 3：width=0，height=1334，size=4.7——width=0，不计算像素密度

Case 4：width=750，height=−1，size=4.7——height<0，不计算像素密度

Case 5：width=750，height=0，size=4.7——height=0，不计算像素密度

Case 6：width=750，height=1334，size=−1——size<0，不计算像素密度

Case 7：width=750，height=1334，size=0——size=0，不计算像素密度

笔者使用的屏幕参数取至 iPhone 6S，iPhone 6S 的屏幕宽为 750 像素，屏幕高为 1334 像素，屏幕尺寸为 4.7 英寸，屏幕像素密度（官网数据）为 326 PPI。官方数据可以作为预期结果，当然，也可以使用上述公式自行计算预期结果。

3.3 设计自动化测试用例

3.3.1 基于 JUnit 设计自动化测试用例

JUnit 作为 Java 单元测试中的首选框架，在 Java 开发中使用最为广泛。目前 JUnit 的最新版本为 JUnit 5，本节内容使用 JUnit 5 作为示例。

1. 配置 JUnit 依赖

在 pom.xml 文件的<name>标签后输入以下粗体部分内容：

```
<project xmlns="http://maven.apache.org/POM/4.0.0"
    xmlns:xsi="http://www.w3.org/2001/XMLSchema-instance"
    xsi:schemaLocation="http://maven.apache.org/POM/4.0.0
http://maven.apache.org/xsd/maven-4.0.0.xsd">
    <modelVersion>4.0.0</modelVersion>
    <groupId>com.lujiatao</groupId>
    <artifactId>unittest</artifactId>
    <version>0.0.1-SNAPSHOT</version>
    <name>Unit Test</name>

    <dependencies>
        <dependency>
            <groupId>org.junit.jupiter</groupId>
            <artifactId>junit-jupiter-api</artifactId>
            <version>5.5.2</version>
            <scope>test</scope>
        </dependency>
        <dependency>
            <groupId>org.junit.jupiter</groupId>
            <artifactId>junit-jupiter-engine</artifactId>
            <version>5.5.2</version>
            <scope>test</scope>
        </dependency>
    </dependencies>

</project>
```

保存 pom.xml 文件，这时 Maven 会自动下载 JUnit 及其依赖的其他 jar 包。

2. 编写自动化测试用例

在 src/test/java 的 com.lujiatao.unittest Package 中创建名为 CalculatorForPpiTest 的 Class，在

CalculatorForPpiTest 中输入以下内容：

```
package com.lujiatao.unittest;

import org.junit.jupiter.api.Assertions;
import org.junit.jupiter.api.BeforeAll;
import org.junit.jupiter.api.Test;

public class CalculatorForPpiTest {
    private static int width;
    private static int height;
    private static double size;

    @BeforeAll
    static void init() {
        width = 750;
        height = 1334;
        size = 4.7;
    }

    @Test
    void testCase1() {
        Assertions.assertEquals(326, CalculatorForPpi.calculate(width, height,
            size));
    }

    @Test
    void testCase2() {
        Assertions.assertEquals(-1, CalculatorForPpi.calculate(-1, height, size));
    }

    @Test
    void testCase3() {
        Assertions.assertEquals(-1, CalculatorForPpi.calculate(0, height, size));
    }

    @Test
    void testCase4() {
        Assertions.assertEquals(-1, CalculatorForPpi.calculate(width, -1, size));
    }

    @Test
    void testCase5() {
        Assertions.assertEquals(-1, CalculatorForPpi.calculate(width, 0, size));
    }

    @Test
    void testCase6() {
        Assertions.assertEquals(-1, CalculatorForPpi.calculate(width, height, -1));
    }
```

```
    @Test
    void testCase7() {
        Assertions.assertEquals(-1, CalculatorForPpi.calculate(width, height, 0));
    }

}
```

保存代码，在 CalculatorForPpiTest.java 上用鼠标右击，从弹出的快捷菜单中选择“Run As → JUnit Test” 选项，运行结果如图 3-1 所示。

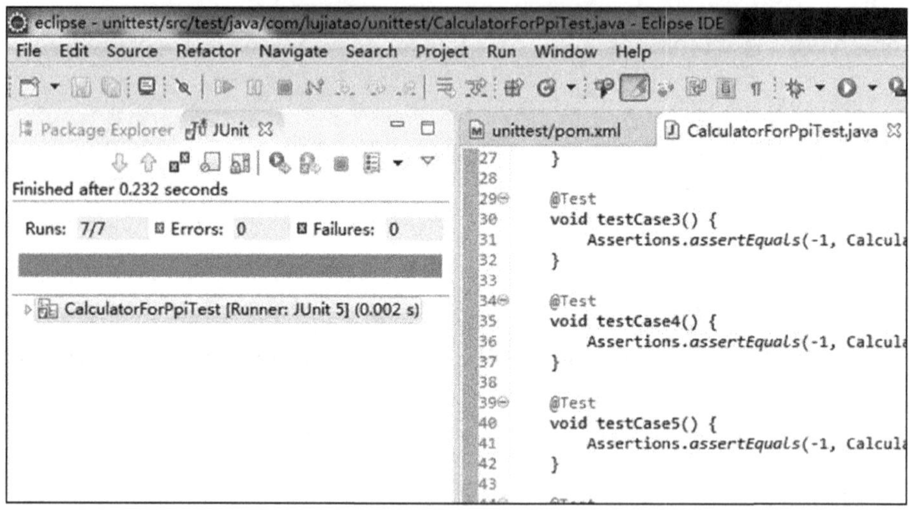

图 3-1

下面对运行结果进行说明。

① init 方法是被@BeforeAll 修饰的，因此是整个 Class 的初始化操作（前置操作）。

② JUnit 使用@Test 注解修饰一个测试用例。

③ 自动化测试的精髓在于可以自动判断用例执行结果，在单元测试中这个过程叫作断言。assertEquals 是 JUnit 中众多断言方法之一，该方法的第一个参数为预期结果，第二个参数为实际结果。

④ 从 JUnit 的运行结果可以看出 7 条测试用例都执行通过了，符合预期。

3.3.2　基于 TestNG 设计自动化测试用例

首先配置 TestNG 依赖，然后在工程（unittest）上用鼠标右击，从弹出的快捷菜单中选择 “TestNG → Convert to TestNG” 选项，在工程中生成 testng.xml 文件。

1. TestNG 断言

TestNG 的断言方法有很多，常用的如下。

assertEquals(String actual, String expected)

assertEquals(String actual, String expected, String message)

判断两个字符串是否相等，如果带第三个参数，那么当断言失败时将显示失败信息，信息的内容即 message。

assertEquals(boolean actual, boolean expected)

assertEquals(boolean actual, boolean expected, String message)

判断两个布尔变量值是否相等，如果带第三个参数，那么当断言失败时将显示失败信息，信息的内容即 message。

assertEquals(int actual, int expected)

assertEquals(int actual, int expected, String message)

判断两个整型变量值是否相等，如果带第三个参数，那么当在断言失败时将显示失败信息，信息的内容即 message。

assertEquals(double actual, double expected, double delta)

assertEquals(double actual, double expected, double delta, String message)

判断两个双精度浮点型变量值是否相等，第三个参数为精度，如果写 0.0，则代表精确到小数点后一位。如果带第四个参数，那么当断言失败时将显示失败信息，信息的内容即 message。

以上断言方法均有对应的“反向”断言方法，即 assertNotEquals，下面举例说明。

删除 CalculatorForPpiTest 中的内容，输入以下代码：

```
package com.lujiatao.unittest;

import org.testng.Assert;
import org.testng.annotations.Test;

public class CalculatorForPpiTest {
    String str1 = "Hello";
    String str2 = "World";

    @Test
    public void testCase1() {
        Assert.assertEquals(str1, str2, "两个字符串不相等！");
    }

    @Test
```

```
    public void testCase2() {
        Assert.assertNotEquals(str1, str2, "两个字符串相等！");
    }

}
```

保存代码，在 testng.xml 上用鼠标右击，从弹出的快捷菜单中选择“Run As → TestNG Suite”选项，此时 Eclipse 的控制台输出如下：

```
[RemoteTestNG] detected TestNG version 6.14.3

===============================================
Suite
Total tests run: 2, Failures: 1, Skips: 0
===============================================
```

从输出结果可以看到，其中一个用例执行失败了，因为 testCase1 断言两个字符串相等，但实际上两个字符串并不相等，所以断言失败。

当断言失败时，在测试报告中可以看到失败信息，如图 3-2 所示。

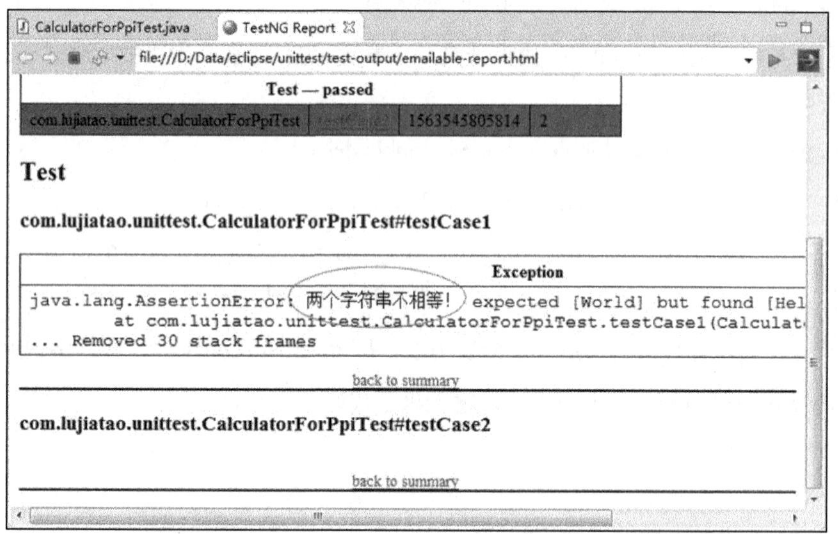

图 3-2

2. 编写自动化测试用例

下面使用 TestNG 代替 JUnit 编写测试用例，代码如下所示：

```
package com.lujiatao.unittest;

import org.testng.Assert;
import org.testng.annotations.BeforeClass;
import org.testng.annotations.Test;
```

```
public class CalculatorForPpiTest {
    private int width;
    private int height;
    private double size;

    @BeforeClass
    public void init() {
        width = 750;
        height = 1334;
        size = 4.7;
    }

    @Test
    public void testCase1() {
        Assert.assertEquals(326, CalculatorForPpi.calculate(width, height, size));
    }

    @Test
    public void testCase2() {
        Assert.assertEquals(-1, CalculatorForPpi.calculate(-1, height, size));
    }

    @Test
    public void testCase3() {
        Assert.assertEquals(-1, CalculatorForPpi.calculate(0, height, size));
    }

    @Test
    public void testCase4() {
        Assert.assertEquals(-1, CalculatorForPpi.calculate(width, -1, size));
    }

    @Test
    public void testCase5() {
        Assert.assertEquals(-1, CalculatorForPpi.calculate(width, 0, size));
    }

    @Test
    public void testCase6() {
        Assert.assertEquals(-1, CalculatorForPpi.calculate(width, height, -1));
    }

    @Test
    public void testCase7() {
        Assert.assertEquals(-1, CalculatorForPpi.calculate(width, height, 0));
    }

}
```

保存代码，在 testng.xml 上用鼠标右击，从弹出的快捷菜单中选择“Run As → TestNG Suite”

选项，然后查看测试报告，如图 3-3 所示。

CalculatorForPpiTest.java　TestNG Report

file:///D:/Data/eclipse/unittest/test-output/emailable-report.html

Test	# Passed	# Skipped	# Failed	Time (ms)	Included Groups	Excluded Groups
Suite						
Test	7	0	0	61		

Class	Method	Start	Time (ms)
Suite			
Test — passed			
com.lujiatao.unittest.CalculatorForPpiTest	[illegible]	1563546069348	9
	[illegible]	1563546069360	2
	[illegible]	1563546069364	2
	[illegible]	1563546069368	2
	[illegible]	1563546069372	1
	[illegible]	1563546069375	2
	[illegible]	1563546069379	1

图 3-3

可以看到与 JUnit 运行结果一致。

在实际项目中，单元自动化测试用例通常是由开发人员编写的，在代码提交到 SVN 或 Git 后，会触发 CI 工具（比如 Jenkins）自动执行单元测试用例，并反馈执行结果。

3.4 Spring 的单元自动化测试

3.4.1 Java 企业级应用简介

在本书中，Spring 相关的示例均采用 Spring Boot 框架。

下面介绍 Java 企业级应用的分层，如图 3-4 所示。

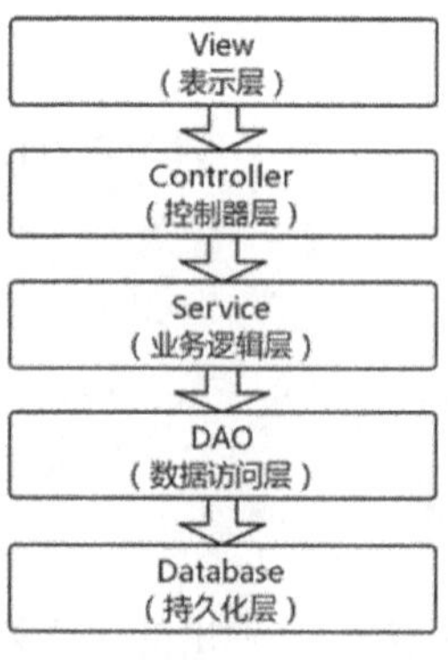

图 3-4

1．表示层

表示层负责和用户进行交互，在 Web 应用中，表示层由 HTML、CSS 和 JavaScript 等组成。现在，前端开发大多使用框架，而不是完全由纯粹的 HTML、CSS 和 JavaScript 实现，常见的前端框架有 jQuery、Bootstrap、Vue.js 和 Angular 等。

2．控制器层

控制器层用来接收用户请求，并转发给业务逻辑的处理组件，即转发给业务逻辑层。在 Java 企业级应用开发中，一般由 Spring MVC 框架充当该角色。

3．业务逻辑层

业务逻辑层是用来实现业务逻辑的，包含 Service、ServiceImpl 和 VO（Value Object，值对象）等。一般来说，Service 为接口，ServiceImpl 为接口的具体实现类，VO 用来在业务逻辑层进行数据传递。在 Java 企业级应用开发中，一般由 Spring 框架充当该角色。

4．数据访问层

数据访问层负责与数据库打交道，包含数据访问对象（Data Access Object，DAO）、DAOImpl 和持久化对象（Persistent Object，PO）等。一般来说，数据访问对象为接口， DAOImpl 为接口的具体实现类。持久化对象为对象关系映射（Object Relational Mapping，ORM），一个 PO 对象代表数据库的一条表记录，PO 对象的属性代表数据库的表字段。数据访问层的框架较多，包括 Spring Data、MyBatis 和 Hibernate 等。

5．持久化层

持久化层的作用是将数据存储在磁盘上。一般由数据库充当该角色，比如 MySQL、SQL Server 或 Oracle 等。

在 Java 企业级应用的整合开发中，有两个经典的框架组合，SSH 和 SSM，它们分别用了不同的框架作为控制器层、业务逻辑层和数据访问层的实现，如表 3-1 所示。

表 3-1

框架组合	Controller	Service	DAO
SSH	Struts	Spring	Hibernate
SSM	Spring MVC	Spring	MyBatis

SSH 框架组合目前已没落，SSM 框架是当前主流框架组合。

说到这里又不得不介绍另一个重要概念——模板引擎（Template Engine）。模板引擎的作用是将用户界面和业务数据分离。常用的 Java 模板引擎有 JSP、FreeMarker、Velocity 和 Thymeleaf

等，Spring Boot 默认集成 Thymeleaf 作为模板引擎。

3.4.2 编写待测程序

1. 工程创建

创建一个新的 Maven 项目，关键信息填写如图 3-5 所示。

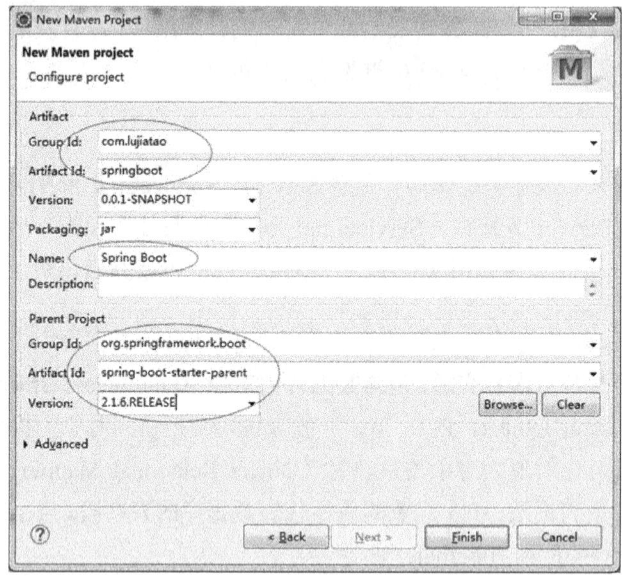

图 3-5

创建完成后，在 pom.xml 文件的<name>标签后输入以下粗体部分内容：

```
<project xmlns="http://maven.apache.org/POM/4.0.0"
    xmlns:xsi="http://www.w3.org/2001/XMLSchema-instance"
    xsi:schemaLocation="http://maven.apache.org/POM/4.0.0
        http://maven.apache.org/xsd/maven-4.0.0.xsd">
    <modelVersion>4.0.0</modelVersion>

    <parent>
        <groupId>org.springframework.boot</groupId>
        <artifactId>spring-boot-starter-parent</artifactId>
        <version>2.1.6.RELEASE</version>
    </parent>

    <groupId>com.lujiatao</groupId>
    <artifactId>springboot</artifactId>
    <version>0.0.1-SNAPSHOT</version>
    <name>Spring Boot</name>

    <dependencies>
```

```
        <dependency>
            <groupId>org.springframework.boot</groupId>
            <artifactId>spring-boot-starter-web</artifactId>
        </dependency>
        <dependency>
            <groupId>org.springframework.boot</groupId>
            <artifactId>spring-boot-starter-thymeleaf</artifactId>
        </dependency>
        <dependency>
            <groupId>org.springframework.boot</groupId>
            <artifactId>spring-boot-starter-test</artifactId>
            <scope>test</scope>
        </dependency>
    </dependencies>

</project>
```

这里添加了 3 个依赖，其中 spring-boot-starter-web 是 Web 开发的基础；spring-boot-starter-thymeleaf 是模板引擎；spring-boot-starter-test 是测试组件，可用于单元自动化测试。保存 pom.xml 文件，这时 Maven 会自动下载 Spring Boot 及其依赖的其他 jar 包。需要说明的是，Spring Boot 项目的依赖较多，下载时间较长，请耐心等待。

2. 编写待测程序

下面用 Web 应用的方式实现本章开头的像素密度计算器。出于个人习惯，笔者喜欢先开发后端，再开发前端。

（1）持久化层实现和数据访问层实现

本程序不需要将数据持久化，因此既不需要实现持久化层，也不需要实现数据访问层。

（2）业务逻辑层实现

在 src/main/java 中创建名为 com.lujiatao.springboot.service 的 Package 及名为 CalculatorForPpiService 的 Class，在 CalculatorForPpiService 中输入以下内容：

```
package com.lujiatao.springboot.service;

import org.springframework.stereotype.Service;

@Service
public class CalculatorForPpiService {

    public long calculate(int width, int height, double size) {
        long result;
        if (width > 0 && height > 0 && size > 0) {
            result = Math.round(Math.pow((Math.pow(width, 2) + Math.pow(height,
               2)) / Math.pow(size, 2), 0.5));
        } else {
            result = -1;
```

```
        }
        return result;
    }

}
```

@Service 是 Spring 框架的业务逻辑层注解。calculate 方法与之前的保持一致。

（3）控制器层实现

在 src/main/java 中创建名为 com.lujiatao.springboot.controller 的 Package 及名为 IndexController 和 CalculatorForPpiController 的 Class，在 IndexController 和 CalculatorForPpiController 中分别输入内容。

IndexController 中的代码如下：

```
package com.lujiatao.springboot.controller;

import org.springframework.stereotype.Controller;
import org.springframework.web.bind.annotation.RequestMapping;

@Controller
public class IndexController {

    @RequestMapping("/index")
    public String index() {
        return "index-page";
    }

}
```

@Controller 是 Spring 框架的控制器层注解，另外，常用的 Spring 框架的控制器层注解还有@RestController。

@RequestMapping 表示接收用户请求。此处代表当用户请求/index 时，使用 index 方法处理。index 方法中返回的“index-page”代表名为“index-page”的页面。

CalculatorForPpiController 中的代码如下：

```
package com.lujiatao.springboot.controller;

import javax.annotation.Resource;

import org.springframework.web.bind.annotation.RequestMapping;
import org.springframework.web.bind.annotation.RequestParam;
import org.springframework.web.bind.annotation.RestController;

import com.lujiatao.springboot.service.CalculatorForPpiService;

@RestController
public class CalculatorForPpiController {
```

```
    @Resource
    private CalculatorForPpiService calculatorForPpiService;

    @RequestMapping("/calculate")
    public String calculate(@RequestParam("width") int width,
       @RequestParam("height") int height,
             @RequestParam("size") double size) {
        long result = calculatorForPpiService.calculate(width, height, size);
        return "{\"PPI\":" + result + "}";
    }

}
```

@Resource 是一种注入注解，将一个 Bean 注入当前类中，即可使用该 Bean。@RequestParam 注解用于接收请求参数，这里使用了 3 次，分别接收前端传入的 with、height 和 size 参数。最后 calculator()方法返回了一个 JSON 格式的字符串，字符串中的反斜杠“\”代表转义。

（4）表示层实现

在 src/main/resources 中创建名为 templates 的目录，在 templates 目录中创建 index-page.html 文件，文件内容如下：

```
<!DOCTYPE html>
<html xmlns:th="http://www.thymeleaf.org">
<head>
<script
    src="https://cdnjs.cloudflare.com/ajax/libs/jquery/3.4.1/jquery.min.js"></s
cript>
<meta charset="UTF-8">
<title>CalculatorForPpi</title>
</head>
<body>
    <form>
        <input type="text" id="width" placeholder="屏幕宽（像素）"><br>
        <input type="text" id="height" placeholder="屏幕高（像素）"><br>
        <input type="text" id="size" placeholder="屏幕尺寸（英寸）"><br>
        <button type="button" id="submit">计算</button>
        <br> <label id="result"></label>
    </form>
    <script>
        $("#submit").click(function() {
            $.ajax({
                url : "/calculate",
                type : "post",
                data : {
                    "width" : $("#width").val(),
                    "height" : $("#height").val(),
                    "size" : $("#size").val(),
                },
                success : function(data) {
```

```
                        let json = JSON.parse(data);
                        $("#result").text("屏幕像素密度为：" + json["PPI"] + " PPI");
                    }
                });
            });
        </script>
    </body>
    </html>
```

<html>标签中的 xmlns 属性表明需要用 Thymeleaf 作为模板引擎。<head>标签中使用 CDN 方式引入 jQuery。<body>中有一个表单，该表单包含 3 个输入框、1 个按钮和 1 个标签。JavaScript 脚本通过 AJAX 发送异步请求传递屏幕的宽、高和尺寸，请求地址为/calculate，请求方式为 POST。后端返回结果后首先解析返回的数据并提取内容，然后将内容显示在表单的标签中，此过程是一个局部刷新页面的过程。

（5）程序入口实现

上述步骤已经实现了程序的功能，但此时程序并没有运行入口，因此无法运行。在 src/main/java 中创建名为 com.lujiatao.springboot 的 Package 及名为 Application 的 Class，在 Application 中输入以下内容：

```
package com.lujiatao.springboot;

import org.springframework.boot.SpringApplication;
import org.springframework.boot.autoconfigure.SpringBootApplication;

@SpringBootApplication
public class Application {

    public static void main(String[] args) {
        SpringApplication.run(Application.class, args);
    }

}
```

@SpringBootApplication 注解代表将程序作为 Spring Boot 应用来运行。调用 SpringApplication 中的 run 方法，并传入当前 Class 作为参数来运行程序，该方法同时传入了 main()方法中的 args 参数。

保存所做的修改，按快捷键 F11 运行工程，此时 Eclipse 的控制台输出如下：

```
  .   ____          _            __ _ _
 /\\ / ___'_ __ _ _(_)_ __  __ _ \ \ \ \
( ( )\___ | '_ | '_| | '_ \/ _` | \ \ \ \
 \\/  ___)| |_)| | | | | || (_| |  ) ) ) )
  '  |____| .__|_| |_|_| |_\__, | / / / /
 =========|_|==============|___/=/_/_/_/
 :: Spring Boot ::        (v2.1.6.RELEASE)
```

```
2019-07-21 21:20:31.831  INFO 8824 --- [           main]
com.lujiatao.springboot.Application      : Starting Application on lujiatao-PC with
PID 8824 (D:\Data\eclipse\springboot\target\classes started by lujiatao in
D:\Data\eclipse\springboot)
2019-07-21 21:20:31.849  INFO 8824 --- [           main]
com.lujiatao.springboot.Application      : No active profile set, falling back to
default profiles: default
2019-07-21 21:20:34.576  INFO 8824 --- [           main]
o.s.b.w.embedded.tomcat.TomcatWebServer  : Tomcat initialized with port(s): 8080
(http)
2019-07-21 21:20:34.621  INFO 8824 --- [           main]
o.apache.catalina.core.StandardService   : Starting service [Tomcat]
2019-07-21 21:20:34.622  INFO 8824 --- [           main]
org.apache.catalina.core.StandardEngine  : Starting Servlet engine: [Apache
Tomcat/9.0.21]
2019-07-21 21:20:34.846  INFO 8824 --- [           main]
o.a.c.c.C.[Tomcat].[localhost].[/]       : Initializing Spring embedded
WebApplicationContext
2019-07-21 21:20:34.846  INFO 8824 --- [           main]
o.s.web.context.ContextLoader            : Root WebApplicationContext:
initialization completed in 2833 ms
2019-07-21 21:20:35.416  INFO 8824 --- [           main]
o.s.s.concurrent.ThreadPoolTaskExecutor  : Initializing ExecutorService
'applicationTaskExecutor'
2019-07-21 21:20:36.178  INFO 8824 --- [           main]
o.s.b.w.embedded.tomcat.TomcatWebServer  : Tomcat started on port(s): 8080 (http)
with context path ''
2019-07-21 21:20:36.183  INFO 8824 --- [           main]
com.lujiatao.springboot.Application      : Started Application in 5.272 seconds (JVM
running for 6.322)
2019-07-21 21:20:55.199  INFO 8824 --- [nio-8080-exec-1]
o.a.c.c.C.[Tomcat].[localhost].[/]       : Initializing Spring DispatcherServlet
'dispatcherServlet'
2019-07-21 21:20:55.199  INFO 8824 --- [nio-8080-exec-1]
o.s.web.servlet.DispatcherServlet        : Initializing Servlet 'dispatcherServlet'
2019-07-21 21:20:55.213  INFO 8824 --- [nio-8080-exec-1]
o.s.web.servlet.DispatcherServlet        : Completed initialization in 14 ms
```

可以看到，Spring Boot 默认使用 Tomcat 作为 Web 服务器，并默认使用 8080 作为端口号。打开浏览器，访问 http://localhost:8080/index，如图 3-6 所示。

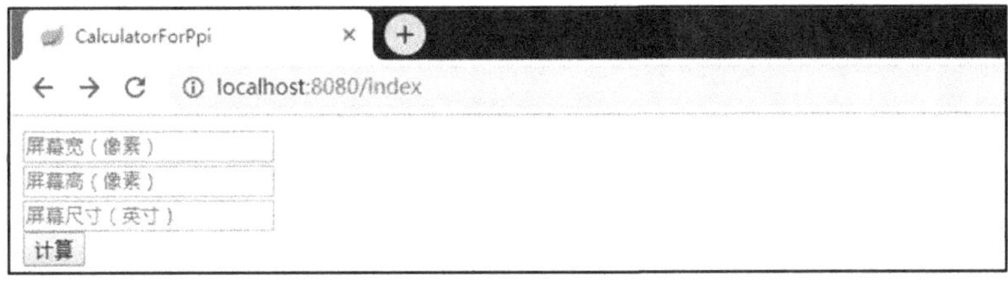

图 3-6

在输入框依次输入 750、1334 和 4.7，单击“计算”按钮，计算结果如图 3-7 所示。

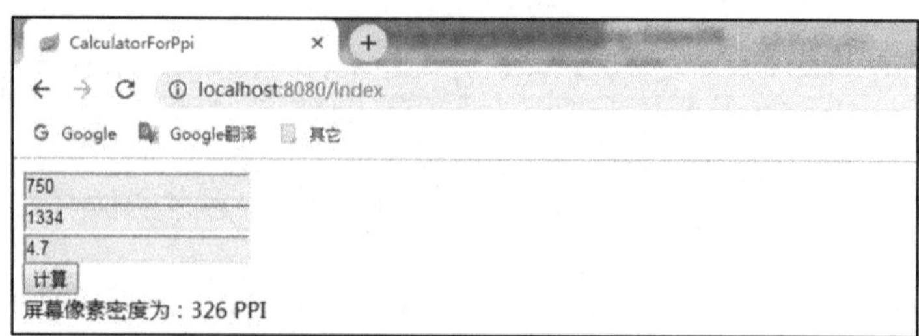

图 3-7

至此，用 Web 应用的方式实现本章开头的像素密度计算器就开发完成了。

3.4.3 单元自动化测试

在此之前需要配置 TestNG 依赖，然后在工程（springboot）上用鼠标右击，从弹出的快捷菜单中选择“TestNG → Convert to TestNG”选项，在工程中生成 testng.xml 文件。

在 Spring 项目中，一般会对业务逻辑层（Service）和控制器层（Controller）进行单元自动化测试，对数据访问层（DAO）进行单元自动化测试的相对较少，因此本章不做介绍。

1. 业务逻辑层单元自动化测试

在 src/test/java 中创建名为 com.lujiatao.springboot.service 的 Package 及名为 CalculatorForPpiServiceTest 的 Class，在 CalculatorForPpiServiceTest 中输入以下内容：

```
package com.lujiatao.springboot.service;

import org.springframework.beans.factory.annotation.Autowired;
import org.springframework.boot.test.context.SpringBootTest;
import org.springframework.test.context.testng.AbstractTestNGSpringContextTests;
import org.testng.Assert;
import org.testng.annotations.BeforeClass;
import org.testng.annotations.Test;

import com.lujiatao.springboot.Application;

@SpringBootTest(classes = { Application.class })
public class CalculatorForPpiServiceTest extends AbstractTestNGSpringContextTests
{
    @Autowired
    private CalculatorForPpiService service;
    private int width;
    private int height;
    private double size;

    @BeforeClass
    public void init() {
        width = 750;
```

```
        height = 1334;
        size = 4.7;
    }

    @Test
    public void testCase1() {
        Assert.assertEquals(326, service.calculate(width, height, size));
    }

    @Test
    public void testCase2() {
        Assert.assertEquals(-1, service.calculate(-1, height, size));
    }

    @Test
    public void testCase3() {
        Assert.assertEquals(-1, service.calculate(0, height, size));
    }

    @Test
    public void testCase4() {
        Assert.assertEquals(-1, service.calculate(width, -1, size));
    }

    @Test
    public void testCase5() {
        Assert.assertEquals(-1, service.calculate(width, 0, size));
    }

    @Test
    public void testCase6() {
        Assert.assertEquals(-1, service.calculate(width, height, -1));
    }

    @Test
    public void testCase7() {
        Assert.assertEquals(-1, service.calculate(width, height, 0));
    }

}
```

下面对代码进行说明。

① @SpringBootTest 注解为 Spring Boot 提供的专为测试设计的注解，classes 方法的作用是传入启动类。

② AbstractTestNGSpringContextTests 是一个抽象类，是 Spring Boot 专为 TestNG 打造的。

③ @Autowired 注解的作用类似于@Resource 注解，不同的是，@Autowired 注解属于 Spring 框架，而@Resource 注解不属于 Spring 框架。

④ 测试用例的写法和之前保持一致，此处不再赘述。

在 tesng.xml 文件的<test>标签中输入以下粗体部分的代码：

```
<?xml version="1.0" encoding="UTF-8"?>
<!DOCTYPE suite SYSTEM "http://testng.org/testng-1.0.dtd">
<suite name="Suite">
    <test thread-count="5" name="Test">
        <classes>
            <class
                name="com.lujiatao.springboot.service.
CalculatorForPpiServiceTest" />
        </classes>
    </test> <!-- Test -->
</suite> <!-- Suite -->
```

保存所做的修改，在 testng.xml 上用鼠标右击，从弹出的快捷菜单中选择“Run As → TestNG Suite”选项，查看测试报告，如图 3-8 所示。

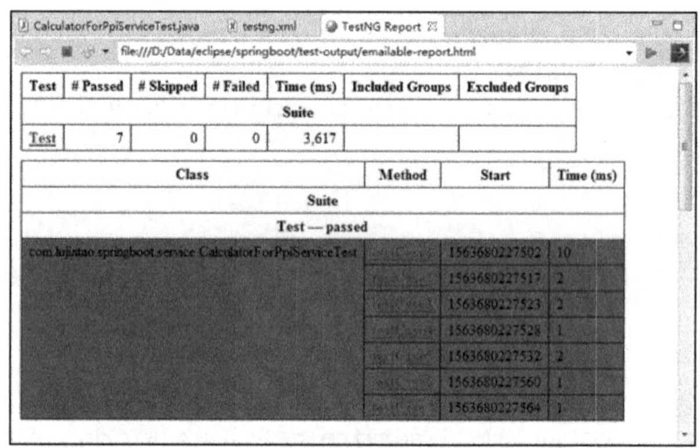

图 3-8

可以看到，运行结果与预期相符。

2. 控制器层单元自动化测试

在 src/test/java 中创建名为 com.lujiatao.springboot.controller 的 Package 及名为 CalculatorForPpiControllerTest 的 Class，在 CalculatorForPpiControllerTest 中输入以下内容：

```
package com.lujiatao.springboot.controller;

import org.springframework.beans.factory.annotation.Autowired;
import org.springframework.boot.test.context.SpringBootTest;
import org.springframework.test.context.testng.AbstractTestNGSpringContextTests;
import org.springframework.test.web.servlet.MockMvc;
import org.springframework.test.web.servlet.RequestBuilder;
import org.springframework.test.web.servlet.request.MockMvcRequestBuilders;
import org.springframework.test.web.servlet.result.MockMvcResultMatchers;
import org.springframework.test.web.servlet.setup.MockMvcBuilders;
```

```
import org.testng.annotations.BeforeClass;
import org.testng.annotations.Test;

import com.lujiatao.springboot.Application;

@SpringBootTest(classes = { Application.class })
public class CalculatorForPpiControllerTest extends
AbstractTestNGSpringContextTests {
    @Autowired
    private CalculatorForPpiController controller;
    private MockMvc mock;
    private RequestBuilder request;
    private String width;
    private String height;
    private String size;

    @BeforeClass
    public void init() {
        mock = MockMvcBuilders.standaloneSetup(controller).build();
        width = "750";
        height = "1334";
        size = "4.7";
    }

    @Test
    public void testCase1() {
        sendRequest(width, height, size, "326");
    }

    @Test
    public void testCase2() {
        sendRequest("-1", height, size, "-1");
    }

    @Test
    public void testCase3() {
        sendRequest("0", height, size, "-1");
    }

    @Test
    public void testCase4() {
        sendRequest(width, "-1", size, "-1");
    }

    @Test
    public void testCase5() {
        sendRequest(width, "0", size, "-1");
    }

    @Test
    public void testCase6() {
```

```
        sendRequest(width, height, "-1", "-1");
    }

    @Test
    public void testCase7() {
        sendRequest(width, height, "0", "-1");
    }

    private void sendRequest(String width, String height, String size, String
      expected) {
        request = MockMvcRequestBuilders.post("/calculate").param("width",
            width).param("height", height).param("size",
                size);
        try {

    mock.perform(request).andExpect(MockMvcResultMatchers.jsonPath("PPI").
        value(expected));
        } catch (Exception e) {
            e.printStackTrace();
        }
    }

}
```

下面对代码进行说明。

① 使用@Autowired 注解将要使用的 Controller 注入当前类。这里使用的是 Spring 的 Mock 类 MockMvc。控制器层的单元测试需要用 MockMvc 模拟请求的调用。

② 在 init()方法中使用注入的 Controller 创建一个 MockMvc 实例，这里使用的方法是 standaloneSetup()，这种测试方式为独立测试方式。另外，也可以使用 webAppContextSetup()方法，这种测试方式为集成 Web 测试方式。

③ 这里将 7 条测试用例的逻辑全部封装在了 sendRequest 方法中，每条测试用例只需传递参数给 sendRequest 方法即可。在 sendRequest 方法中，首先通过 MockMvcRequestBuilders 构建一个请求对象，然后使用 MockMvc 对象调用该请求，接着使用 andExpect()方法添加断言。在 andExpect()方法中，通过提取后端返回的 JSON 与预期结果进行比较。

④ 在请求的调用和断言过程中可能会抛出异常，因此需要对异常进行处理。这里对异常处理的方式是捕获异常。

修改 tesng.xml 文件，修改以下粗体部分内容：

```
<?xml version="1.0" encoding="UTF-8"?>
<!DOCTYPE suite SYSTEM "http://testng.org/testng-1.0.dtd">
<suite name="Suite">
    <test thread-count="5" name="Test">
        <classes>
            <class
```

```
    name="com.lujiatao.springboot.controller.CalculatorForPpiControllerTest" />
        </classes>
    </test> <!-- Test -->
</suite> <!-- Suite -->
```

保存所做的修改，在 testng.xml 上用鼠标右击，从弹出的快捷菜单中选择“Run As → TestNG Suite”选项，查看测试报告，如图 3-9 所示。

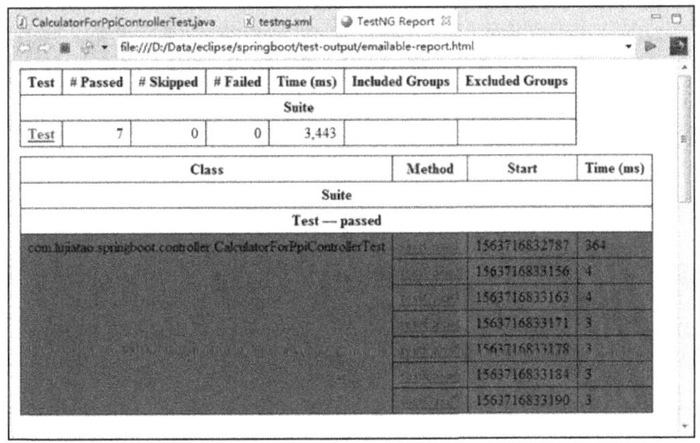

图 3-9

测试用例全部执行通过，运行结果与预期相符。

第4章 HTTP接口自动化测试

4.1 HTTP简介

在 TCP/IP 中，HTTP 属于传输层协议。HTTP 采用“请求—应答”模式，并且该协议是无状态的，即后续的处理如果需要用到前面的信息，则必须重传。HTTP 通过 SSL/TLS 加密成为 HTTPS，与 HTTP 相比，HTTPS 的安全性更好，但传输速度不及 HTTP。

HTTP 中有多种请求方法。

1. GET 方法

GET 方法用于获取指定资源，可理解为“读取”资源。在 GET 方法的 URL 中可以携带参数，携带参数的格式为“key1=value1&key2=value2&key3=value3”。

2. HEAD 方法

与 GET 方法一样，HEAD 方法也用于获取指定资源，区别在于，HEAD 方法的响应报文没有消息体。

3. POST 方法

POST 方法用于创建或修改指定资源，比如常见的提交表单或上传文件等。POST 方法既可以在 URL 中携带参数，也可以在请求体中携带参数。POST 方法和 GET 方法是最常用的两种 HTTP 请求方法。

4. PUT 方法

与 POST 方法一样，PUT 方法也用于创建或修改指定资源，区别在于，PUT 方法是幂等的，即调用一次与调用多次的效果一样；而 POST 方法是非幂等的，即调用多次效果可能有差异。

5. DELETE 方法

DELETE 方法用于请求服务器删除指定的资源。

6. TRACE 方法

TRACE 方法主要用于调试或测试，是对服务器的一种连通性测试方法。

7. OPTIONS 方法

OPTIONS 方法一般用于检测服务器支持的请求方法，响应报文中包含一个名为 Allow 的响应头字段，该字段的值表示了服务器支持的 HTTP 方法。

8. CONNECT 方法

CONNECT 方法一般用于代理服务器。比如，服务器使用 HTTPS 进行数据传输，且浏览器需要代理服务器，那么浏览器首先使用 CONNECT 方法以明文方式向代理服务器发送目标服务器的 IP 地址和端口，在代理服务器与目标服务器建立连接后再进行后续的数据传输。这样做的好处是代理服务器不会破坏 HTTPS 传输过程的安全性。

在客户端发起 HTTP 请求后服务器会进行响应，不同的响应码对应不同的场景。响应码由 3 位数字组成，第一个数字代表当前响应的类型。

1XX——提示信息，服务器的临时响应，此时客户端应继续发起请求。

2XX——成功，请求已被服务器成功处理，比如 200 OK。

3XX——重定向，需要客户端进行后续操作才能达成目的，比如 302 Found。

4XX——客户端错误，客户端请求发生错误，比如 404 Not Found。

5XX——服务器错误，服务器处理正确请求时发生错误，比如 500 Internal Server Error。

4.2　部署待测程序

1. 安装 JDK

在即将部署待测程序的服务器上安装 JDK。

2. 部署待测程序

从笔者的 GitHub 下载待测程序，程序名为 httpinterface-0.0.1-SNAPSHOT.jar，下载后放在服务器上，执行以下命令运行即可：

```
java -jar E:\httpinterface-0.0.1-SNAPSHOT.jar
```

上述命令把本地电脑作为了服务器，且把待测程序放在 E 盘根目录，读者可根据实际情况替换以上路径。运行成功后如图 4-1 所示。

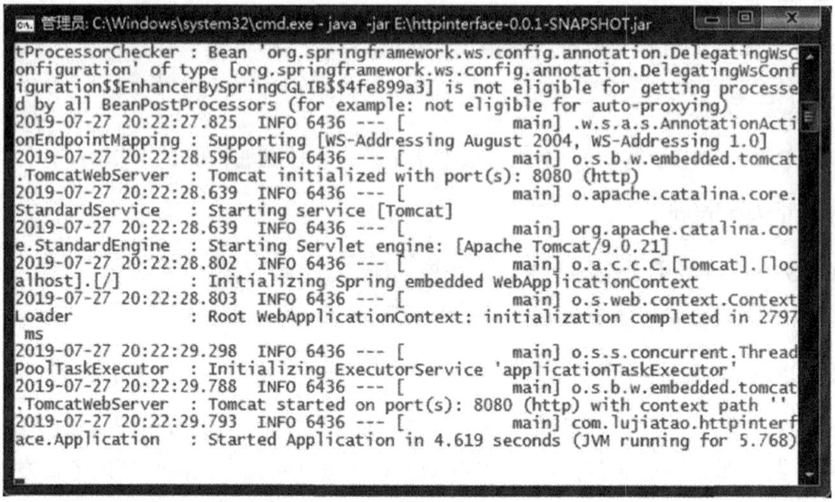

```
管理员: C:\Windows\system32\cmd.exe - java -jar E:\httpinterface-0.0.1-SNAPSHOT.jar
tProcessorChecker : Bean 'org.springframework.ws.config.annotation.DelegatingWsC
onfiguration' of type [org.springframework.ws.config.annotation.DelegatingWsConf
iguration$$EnhancerBySpringCGLIB$$4fe899a3] is not eligible for getting processe
d by all BeanPostProcessors (for example: not eligible for auto-proxying)
2019-07-27 20:22:27.825  INFO 6436 --- [           main] .w.s.a.s.AnnotationActi
onEndpointMapping : Supporting [WS-Addressing August 2004, WS-Addressing 1.0]
2019-07-27 20:22:28.596  INFO 6436 --- [           main] o.s.b.w.embedded.tomcat
.TomcatWebServer  : Tomcat initialized with port(s): 8080 (http)
2019-07-27 20:22:28.639  INFO 6436 --- [           main] o.apache.catalina.core.
StandardService   : Starting service [Tomcat]
2019-07-27 20:22:28.639  INFO 6436 --- [           main] org.apache.catalina.cor
e.StandardEngine  : Starting Servlet engine: [Apache Tomcat/9.0.21]
2019-07-27 20:22:28.802  INFO 6436 --- [           main] o.a.c.c.C.[Tomcat].[loc
alhost].[/]       : Initializing Spring embedded WebApplicationContext
2019-07-27 20:22:28.803  INFO 6436 --- [           main] o.s.web.context.Context
Loader            : Root WebApplicationContext: initialization completed in 2797
 ms
2019-07-27 20:22:29.298  INFO 6436 --- [           main] o.s.s.concurrent.Thread
PoolTaskExecutor  : Initializing ExecutorService 'applicationTaskExecutor'
2019-07-27 20:22:29.788  INFO 6436 --- [           main] o.s.b.w.embedded.tomcat
.TomcatWebServer  : Tomcat started on port(s): 8080 (http) with context path ''
2019-07-27 20:22:29.793  INFO 6436 --- [           main] com.lujiatao.httpinterf
ace.Application   : Started Application in 4.619 seconds (JVM running for 5.768)
```

图 4-1

4.3 手工测试用例设计

4.3.1 分析待测接口

在做接口测试时，需要以接口文档作为标准。接口文档一般以 Word 文档、Excel 表格或接口文档服务器等形式呈现。下面以 Excel 表格形式给出待测程序的接口文档，文档内容如表 4-1 所示。

表 4-1

1. 通过手机型号获取手机信息	
接口类型	RESTful
请求类型	GET
接口路径	/mobilePhone
请求内容示例	model=moto+Z+Play

响应内容示例	{
	"brand": "Motorola",
	"model": "moto Z Play",
	"os": "ANDROID"
	}

续表

2. 保存手机信息	
接口类型	RESTful
请求类型	POST
接口路径	/mobilePhone

请求内容示例	{
	"brand": "Motorola",
	"model": "moto Z Play",
	"os": "ANDROID"
	}

响应内容示例	{
	"code": 0,
	"message": "保存成功！"
	}

3. 通过手机型号获取手机信息	
接口类型	SOAP
请求类型	POST
接口路径	/MobilePhones
请求内容示例	<SOAP-ENV:Envelope xmlns:SOAP-ENV="http://schemas.xmlsoap.org/soap/envelope/" xmlns:hi="http://www.lujiatao.com/httpinterface/MobilePhones"> <SOAP-ENV:Header/> <SOAP-ENV:Body> <hi:getMobilePhoneRequest> <hi:model>iPhone 6S</hi:model> </hi:getMobilePhoneRequest> </SOAP-ENV:Body> </SOAP-ENV:Envelope>
响应内容示例	<SOAP-ENV:Envelope xmlns:SOAP-ENV="http://schemas.xmlsoap.org/soap/envelope/"> <SOAP-ENV:Header/> <SOAP-ENV:Body> <ns2:getMobilePhoneResponse xmlns:ns2="http://www.lujiatao.com/httpinterface/MobilePhones">

续表

响应内容示例	<ns2:mobilePhone> <ns2:brand>Apple</ns2:brand> <ns2:model>iPhone 6S</ns2:model> <ns2:os>IOS</ns2:os> </ns2:mobilePhone> </ns2:getMobilePhoneResponse> </SOAP-ENV:Body> </SOAP-ENV:Envelope>

这是一个简单的接口文档，在实际项目中，接口文档应该更为规范。

1．RESTful 接口分析

待测程序有两个 RESTful 接口。

/mobilePhone GET 接口：入参为手机型号，接口功能为通过手机型号获取手机信息，包括手机品牌、手机型号和手机操作系统。

/mobilePhone POST 接口：入参为手机信息，接口功能为保存手机信息。

2．SOAP 接口分析

待测程序只有一个 SOAP 接口，入参为手机型号，接口功能为通过手机型号获取手机信息，包括手机品牌、手机型号和手机操作系统。

SOAP 接口的功能和上面的/mobilePhone GET 接口的功能是一样的，这样编写待测程序的目的是方便读者比较两种接口的差异。RESTful 接口一般基于 JSON 传输数据，而 SOAP 接口基于 XML 传输数据，因此 RESTful 接口的效率更高，这也是目前 RESTful 接口更为流行的原因之一。

4.3.2 测试用例设计

接口测试常用的工具有 JMeter、Postman 和 SoapUI 等，无论使用哪个工具，都应提前做好接口测试用例设计。

接口测试用例设计应该从入参、返回值、接口逻辑、性能和安全等各个方面考虑。这里以入参作为示例设计手工测试用例，后面会将这部分手工测试用例转换为自动化测试用例。

1．RESTful 接口测试用例设计

（1）/mobilePhone GET 接口

① 参数必填项。

Case 1：model="iPhone 6S"——返回 iPhone 6S 手机（假设存在 iPhone 6S 手机）

Case 2：model=""——返回空

Case 3：model=null——返回空

② 参数长度。

接口文档未提及参数长度， 理论上 GET 请求没有参数长度限制，但我们可以提交一个相对较长的参数。

Case 4：model="01234567890123456789012345678901234567890123456789"——返回空

③ 参数组合。

在多个参数的接口中，参数可能相互响应或制约，因此对参数组合的用例设计必不可少。本接口不涉及（因为只有一个参数）。

④ 参数规则。

一般特殊的参数会有相应规则，比如手机号、身份证号或统一社会信用代码等。本接口不涉及。

⑤ 参数枚举。

有些参数只接收几个固定的参数值，这类参数在服务端大多用枚举定义，这种情况下需要测试每种枚举值。本接口不涉及。

综合以上各种情况，/mobilePhone GET 接口一共设计了 4 条手工测试用例。

（2）/mobilePhone POST 接口

① 参数必填项。

Case 1：{"brand":"Motorola","model":"moto Z Play","os":"ANDROID"}——保存成功

Case 2：{"brand":"","model":"moto Z Play","os":"ANDROID"}——保存失败

Case 3：{"brand":null,"model":"moto Z Play","os":"ANDROID"}——保存失败

Case 4：{"brand":"Motorola","model":"","os":"ANDROID"}——保存失败

Case 5：{"brand":"Motorola","model":null,"os":"ANDROID"}——保存失败

Case 6：{"brand":"Motorola","model":"moto Z Play","os":""}——保存失败

Case 7：{"brand":"Motorola","model":"moto Z Play","os":null}——保存失败

② 参数长度。

接口文档未提及参数长度，理论上 POST 请求没有限制请求体长度，但我们可以提交一个相对较长的参数。

Case 8：

{"brand":"01234567890123456789012345678901234567890123456789","model":"moto Z Play","os":"ANDROID"}——保存成功

Case 9：

{"brand":"Motorola","model":"01234567890123456789012345678901234567890123456789","os":"ANDROID"}——保存成功

这里没有对 os 参数做参数长度测试，稍后会说明原因。

③ 参数组合。

本接口不涉及。

④ 参数规则。

本接口不涉及。

⑤ 参数枚举。

os 参数代表操作系统，目前主流的手机操作系统是 Android 和 iOS。接口文档示例中 os 的值为“ANDROID”，在 Java 中枚举用大写表示，故推测该字段在后台以枚举方式定义。

Case 10：{"brand":"Apple","model":"iPhone XS Max","os":"IOS"}——保存成功

Case 11：{"brand":"Nokia","model":"N95","os":"SYMBIAN"}——保存失败

os 的值为“ANDROID”，上述已有用例覆盖，因此不再单独设计用例进行覆盖。

综合以上各种情况，/mobilePhone POST 接口一共设计了 11 条手工测试用例。

2. SOAP 接口测试用例设计

由于待测程序中 SOAP 接口的功能和上面的/mobilePhone GET 接口的功能是一样的，因此手工测试用例设计参考/mobilePhone GET 接口，不再赘述。

4.4 HttpClient 用法

4.4.1 HttpClient 简介

尽管 JDK 中有通过 HTTP 访问资源的相关 API，但其功能和灵活性受到限制。HttpClient 隶属于 Apache 的 HttpComponents 项目，它提供了更丰富和更高效的 HTTP 访问方式，主要特

点如下。

- 纯 Java 实现，支持 HTTP 1.0 和 1.1。
- 支持 HTTPS。
- 支持通过 CONNECT 方法建立隧道连接。
- 支持 Basic、Digest 和 NTLM 等鉴权方式。
- 支持设置连接超时。
- 源代码基于 Apache 许可协议，因此是免费开源的。

4.4.2 创建工程

在学习 HttpClient 之前，先来创建一个新的 Maven 项目，关键信息填写如下。

Group Id：com.lujiatao

Artifact Id：httpinterfacetest

Name：HTTP Interface Test

可根据实际情况填写，不需要和这里填写的内容完全一致。

创建完成后，在 pom.xml 文件的<name>标签后输入以下粗体部分内容：

```
<project xmlns="http://maven.apache……"
    xmlns:xsi="http://www.w3……"
    xsi:schemaLocation="http://maven.apache…… http://maven.apache……">
    <modelVersion>4.0.0</modelVersion>

    <groupId>com.lujiatao</groupId>
    <artifactId>httpinterfacetest</artifactId>
    <version>0.0.1-SNAPSHOT</version>
    <name>HTTP Interface Test</name>

    <dependencies>
        <dependency>
            <groupId>org.apache.httpcomponents</groupId>
            <artifactId>httpclient</artifactId>
            <version>4.5.9</version>
        </dependency>
    </dependencies>

</project>
```

保存 pom.xml 文件，这时 Maven 会自动下载 HttpClient 及其依赖的其他 jar 包。

4.4.3 发送 HTTP 请求

1. 发送 GET 请求

在 src/test/java 中创建名为 com.lujiatao.httpinterfacetest 的 Package 及名为 Test 的 Class（需要勾选"public static void main(String[] args)"），在 Test 中输入以下内容：

```
package com.lujiatao.httpinterfacetest;

import java.net.URI;
import java.net.URISyntaxException;

import org.apache.http.client.methods.CloseableHttpResponse;
import org.apache.http.client.methods.HttpGet;
import org.apache.http.client.utils.URIBuilder;
import org.apache.http.impl.client.CloseableHttpClient;
import org.apache.http.impl.client.HttpClients;
import org.apache.http.util.EntityUtils;

public class Test {

    public static void main(String[] args) {
        CloseableHttpClient client = HttpClients.createDefault();
        HttpGet httpGet1 = new HttpGet("http://localhost:8080/mobilePhone?model=
            iPhone+6S");
        URI uri = null;
        try {
            uri = new URIBuilder().setScheme("http").setHost("localhost").
                setPort(8080).setPath("/mobilePhone")
                    .setParameter("model", "iPhone 6S").build();
        } catch (URISyntaxException e) {
            e.printStackTrace();
        }
        HttpGet httpGet2 = new HttpGet(uri);
        CloseableHttpResponse response = null;
        try {
            response = client.execute(httpGet1);
            System.out.println(EntityUtils.toString(response.getEntity()));
            response = client.execute(httpGet2);
            System.out.println(EntityUtils.toString(response.getEntity()));
        } catch (Exception e) {
            e.printStackTrace();
        }
    }

}
```

保存代码，按快捷键 F11 运行工程，此时 Eclipse 的控制台输出如下：

```
{"brand":"Apple","model":"iPhone 6S","os":"IOS"}
{"brand":"Apple","model":"iPhone 6S","os":"IOS"}
```

下面对运行结果进行说明。

① 通过调用 HttpClients 类的静态方法构建一个 HTTP 客户端。

② GET 请求通过实例化 HttpGet 类来实现，其他 HTTP 请求也有相应的类，即 HttpDelete、HttpHead、HttpOptions、HttpPatch、HttpPost、HttpPut 和 HttpTrace。

③ 第一种构建 GET 请求的方式是使用 String，其特点是语法简洁，但结构不是很清晰。第二种构建 GET 请求的方式是使用 URI（Uniform Resource Identifier，即统一资源标识符），其特点是结构清晰，但构建过程较为复杂，且构建过程可能导致 URI 语法异常，这里我们对异常进行了捕获和处理。当遇到空格时，第一种方式需要手动编码（即把空格替换成+号），而第二种方式会自动编码。

④ 通过 HTTP 客户端执行请求得到服务器响应，并将响应打印到控制台。

⑤ 发起请求的过程可能导致多种异常，为简明步骤，这里用 Exception 表示，因为 Exception 类为其他几个异常类的父类，这样可避免罗列多个异常类。

2. 发送 POST 请求

下面介绍发送 RESTful POST 请求和 SOAP 请求。SOAP 请求是通过 POST 请求发送的，请求体使用 XML 传输数据。

删除 Test 中的内容，输入以下内容：

```
package com.lujiatao.httpinterfacetest;

import org.apache.http.client.methods.CloseableHttpResponse;
import org.apache.http.client.methods.HttpPost;
import org.apache.http.entity.ContentType;
import org.apache.http.entity.StringEntity;
import org.apache.http.impl.client.CloseableHttpClient;
import org.apache.http.impl.client.HttpClients;
import org.apache.http.util.EntityUtils;

public class Test {

    public static void main(String[] args) {
        CloseableHttpClient client = HttpClients.createDefault();
// 构建和发送 RESTful POST 请求
        HttpPost rest = new HttpPost("http://localhost:8080/mobilePhone");
        StringEntity stringEntity = new StringEntity(
                "{\"brand\":\"Motorola\",\"model\":\"moto Z Play\",\"os\":\
                "ANDROID\"}", ContentType.APPLICATION_JSON);
        rest.setEntity(stringEntity);
        CloseableHttpResponse response = null;
        try {
            response = client.execute(rest);
```

```
            System.out.println(EntityUtils.toString(response.getEntity()));
        } catch (Exception e) {
            e.printStackTrace();
        }
// 构建和发送 SOAP 请求
        HttpPost soap = new HttpPost("http://localhost:8080/MobilePhones");
        String soapString = "<SOAP-ENV:Envelope
           xmlns:SOAP-ENV=\"http://schemas.xmlsoap.org/soap/envelope/\"
           xmlns:hi=\"http://www.lujiatao……\">\r\n"
                + "   <SOAP-ENV:Header/>\r\n" + "   <SOAP-ENV:Body>\r\n" + "
                   <hi:getMobilePhoneRequest>\r\n"
                + "          <hi:model>iPhone 6S</hi:model>\r\n" + "
                   </hi:getMobilePhoneRequest>\r\n"
                + "   </SOAP-ENV:Body>\r\n" + "</SOAP-ENV:Envelope>";
        stringEntity = new StringEntity(soapString, ContentType.TEXT_XML);
        soap.setEntity(stringEntity);
        response = null;
        try {
            response = client.execute(soap);
            System.out.println(EntityUtils.toString(response.getEntity()));
        } catch (Exception e) {
            e.printStackTrace();
        }
    }

}
```

保存代码，按快捷键 F11 运行工程，此时 Eclipse 的控制台输出如下：

```
{"code":0,"message":"保存成功！"}
<SOAP-ENV:Envelope xmlns:SOAP-ENV="http://schemas.xmlsoap……
"><SOAP-ENV:Header/><SOAP-ENV:Body><ns2:getMobilePhoneResponse
xmlns:ns2="http://www.lujiatao……"><ns2:mobilePhone><ns2:brand>Apple
</ns2:brand><ns2:model>iPhone
6S</ns2:model><ns2:os>IOS</ns2:os></ns2:mobilePhone></ns2:getMobilePhoneRespons
e></SOAP-ENV:Body></SOAP-ENV:Envelope>
```

下面对运行结果进行说明。

① 两种请求均使用 String 的方式构建。

② 由于 SOAP 请求的请求体跨越了多行，故通过加号（+）进行连接。

③ 两种请求均通过 StringEntity 对象给请求设置了请求体和 Content-Type，区别在于，RESTful POST 请求 Content-Type 为 APPLICATION_JSON，而 SOAP 请求 Content-Type 为 TEXT_XML。

4.4.4 处理服务器响应

服务器处理完客户端发来的 HTTP 请求后，会将结果发回给客户端。HTTP 响应由响应行、响应头和响应体组成，而 HTTP 请求也包含三个组成部分，即请求行、请求头和请求体。前面只是将服务器响应的实体转换为字符串并打印到控制台，其实对服务器响应还可以做很多事情。

删除 Test 中的内容，输入以下代码：

```
package com.lujiatao.httpinterfacetest;

import org.apache.http.HttpEntity;
import org.apache.http.client.methods.CloseableHttpResponse;
import org.apache.http.client.methods.HttpGet;
import org.apache.http.impl.client.CloseableHttpClient;
import org.apache.http.impl.client.HttpClients;
import org.apache.http.util.EntityUtils;

public class Test {

    public static void main(String[] args) {
        CloseableHttpClient client = HttpClients.createDefault();
        HttpGet httpGet = new
           HttpGet("http://localhost:8080/mobilePhone?model=iPhone+6S");
        CloseableHttpResponse response = null;
        try {
            response = client.execute(httpGet);
            System.out.println(response.getProtocolVersion());
            System.out.println(response.getStatusLine().getStatusCode());
            System.out.println(response.getStatusLine().toString());
            HttpEntity httpEntity = response.getEntity();
            System.out.println(httpEntity.getContentType());
            response.close();
        } catch (Exception e) {
            e.printStackTrace();
        }
    }

}
```

保存代码，按快捷键 F11 运行工程，此时 Eclipse 的控制台输出如下：

```
HTTP/1.1
200
HTTP/1.1 200
Content-Type: application/json;charset=UTF-8
```

下面对运行结果进行说明。

① 通过服务器响应可以得到协议版本和响应行，通过响应行又可以进一步得到响应码。当然，也可以调用 toString()方法将整个响应行转换为字符串。

② 通过调用 getEntity()方法可以得到响应实体，通过响应实体又可以得到 Content-Type。

③ 最后关闭服务器响应，这里调用了 close()方法。

4.4.5 设置请求头

HttpClient 可以设置请求头。在实际项目中，登录后往往需要将 session 放置在 cookie 中，以供后续接口使用，而 cookie 正是请求头的一部分。

删除 Test 中的内容，输入以下代码：

```
package com.lujiatao.httpinterfacetest;

import org.apache.http.Header;
import org.apache.http.client.methods.HttpPost;

public class Test {

    public static void main(String[] args) {
        HttpPost httpPost = new HttpPost("http://ip:port/uri");
        httpPost.addHeader("Cookie", "key1=value1; key2=value2");
        httpPost.addHeader("User-Agent", "My User Agent");
        for (Header header : httpPost.getAllHeaders()) {
            System.out.println(header);
        }
    }

}
```

保存代码，按快捷键 F11 运行工程，此时 Eclipse 的控制台输出如下：

```
Cookie: key1=value1; key2=value2
User-Agent: My User Agent
```

从输出中可以看出，HTTP 请求可以通过 addHeader()方法设置请求头，通过 getAllHeaders()方法获取所有请求头（用一个请求头数组表示）。

4.5 TestNG 集成 HttpClient

首先将 TestNG 和 HttpClient 进行集成，然后进行 HTTP 接口自动化测试。

在 pom.xml 文件的<dependencies>标签中输入以下粗体部分内容：

```
<project xmlns="http://maven.apache……"
    xmlns:xsi="http://www.w3……"
    xsi:schemaLocation="http://maven.apache…… http://maven.apache……">
    <modelVersion>4.0.0</modelVersion>

    <groupId>com.lujiatao</groupId>
    <artifactId>httpinterfacetest</artifactId>
    <version>0.0.1-SNAPSHOT</version>
    <name>HTTP Interface Test</name>
```

```
    <dependencies>
        <dependency>
            <groupId>org.apache.httpcomponents</groupId>
            <artifactId>httpclient</artifactId>
            <version>4.5.9</version>
        </dependency>
        <dependency>
            <groupId>org.testng</groupId>
            <artifactId>testng</artifactId>
            <version>6.14.3</version>
            <scope>test</scope>
        </dependency>
        <dependency>
            <groupId>org.json</groupId>
            <artifactId>json</artifactId>
            <version>20180813</version>
        </dependency>
        <dependency>
            <groupId>org.dom4j</groupId>
            <artifactId>dom4j</artifactId>
            <version>2.1.1</version>
        </dependency>
        <dependency>
            <groupId>jaxen</groupId>
            <artifactId>jaxen</artifactId>
            <version>1.2.0</version>
        </dependency>
    </dependencies>

</project>
```

保存 pom.xml 文件，这时 Maven 会自动下载依赖的 jar 包。这里添加了 4 个依赖。

① TestNG：自动化测试框架 TestNG 的依赖。

② JSON：用于处理 JSON 数据，在 RESTful POST 请求的请求体及响应体中均可使用。

③ DOM4J：用于处理 XML 数据，在 SOAP 请求的请求体及响应体中均可使用。

④ JAXEN：当 DOM4J 用 XPath 方式获取节点时需要依赖该 jar 包。

依赖 jar 包下载完成后，在工程（httpinterfacetest）上用鼠标右击，从弹出的快捷菜单中选择“TestNG → Convert to TestNG” 选项，在工程中生成 testng.xml 文件。

4.5.1　RESTful 接口自动化测试

1．编写 GET 接口自动化测试用例

把 Test Class 重命名为 GetMobilePhoneTest，删除 GetMobilePhoneTest 中的内容，输入以下代码：

```
package com.lujiatao.httpinterfacetest;

import java.net.URI;

import org.apache.http.client.methods.CloseableHttpResponse;
import org.apache.http.client.methods.HttpGet;
import org.apache.http.client.utils.URIBuilder;
import org.apache.http.impl.client.CloseableHttpClient;
import org.apache.http.impl.client.HttpClients;
import org.apache.http.util.EntityUtils;
import org.testng.Assert;
import org.testng.annotations.AfterClass;
import org.testng.annotations.BeforeClass;
import org.testng.annotations.Test;

public class GetMobilePhoneTest {
    private CloseableHttpClient client;
    private CloseableHttpResponse response;

    @BeforeClass
    public void init() {
        client = HttpClients.createDefault();
    }

    @Test
    public void testCase1() {
        Assert.assertEquals("{\"brand\":\"Apple\",\"model\":\"iPhone
           6S\",\"os\":\"IOS\"}",
                sendHttpGetRequest(client, "iPhone 6S"));
    }

    @Test
    public void testCase2() {
        Assert.assertEquals("", sendHttpGetRequest(client, ""));
    }

    @Test
    public void testCase3() {
        Assert.assertEquals("", sendHttpGetRequest(client, null));
    }

    @Test
    public void testCase4() {
        Assert.assertEquals("", sendHttpGetRequest(client,
           "01234567890123456789012345678901234567890123456789"));
    }

    @AfterClass
    public void clear() {
```

```
        try {
            response.close();
            client.close();
        } catch (Exception e) {
            e.printStackTrace();
        }
    }

    private String sendHttpGetRequest(CloseableHttpClient client, String model) {
        String result = null;
        try {
            URI uri = new URIBuilder().setScheme("http").setHost("localhost").
                setPort(8080).setPath("/mobilePhone")
                    .setParameter("model", model).build();
            response = client.execute(new HttpGet(uri));
            result = EntityUtils.toString(response.getEntity());
        } catch (Exception e) {
            e.printStackTrace();
        }
        return result;
    }

}
```

修改 testng.xml 文件，在<test>标签中新增以下粗体部分内容：

```
<?xml version="1.0" encoding="UTF-8"?>
<!DOCTYPE suite SYSTEM "http://testng.org/testng-1.0.dtd">
<suite name="Suite">
    <test thread-count="5" name="Test">
        <classes>
            <class name="com.lujiatao.httpinterfacetest.GetMobilePhoneTest" />
        </classes>
    </test> <!-- Test -->
</suite> <!-- Suite -->
```

保存所做的修改，在 testng.xml 上用鼠标右击，从弹出的快捷菜单中选择"Run As → TestNG Suite"选项，可以看到如图 4-2 所示的测试报告。

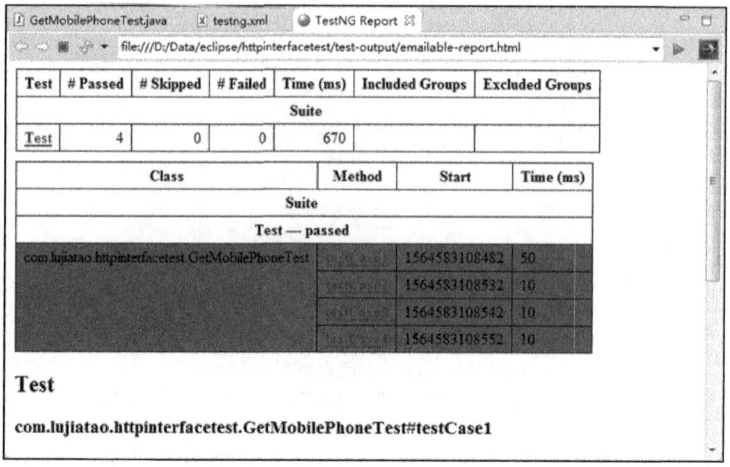

GetMobilePhoneTest.java　testng.xml　TestNG Report

file:///D:/Data/eclipse/httpinterfacetest/test-output/emailable-report.html

Test	# Passed	# Skipped	# Failed	Time (ms)	Included Groups	Excluded Groups
Suite						
Test	4	0	0	670		

Class	Method	Start	Time (ms)
Suite			
Test — passed			
com.lujiatao.httpinterfacetest.GetMobilePhoneTest	[illegible]	1564583108482	50
	[illegible]	1564583108532	10
	[illegible]	1564583108542	10
	[illegible]	1564583108552	10

Test

com.lujiatao.httpinterfacetest.GetMobilePhoneTest#testCase1

图 4-2

下面对运行结果进行说明。

① init()方法初始化了一个 HTTP 客户端，并在 sendHttpGetRequest()方法中使用，最后在 clear()方法中关闭。

② 将构造 URI、发送 GET 请求和转换服务器响应等过程封装在 sendHttpGetRequest()方法中，通过传递 HTTP 客户端和参数来使用该方法。

③ 在 clear()方法中对 HTTP 客户端和服务器响应进行关闭操作，在操作过程中可能出现异常，所以需要对异常进行捕获和处理。

2. POST 接口自动化测试用例编写

在 com.lujiatao.httpinterfacetest Package 中创建名为 SaveMobilePhoneTest 的 Class，在 SaveMobilePhoneTest 中输入以下内容：

```
package com.lujiatao.httpinterfacetest;

import org.apache.http.client.methods.CloseableHttpResponse;
import org.apache.http.client.methods.HttpPost;
import org.apache.http.entity.ContentType;
import org.apache.http.entity.StringEntity;
import org.apache.http.impl.client.CloseableHttpClient;
import org.apache.http.impl.client.HttpClients;
import org.apache.http.util.EntityUtils;
import org.testng.Assert;
import org.testng.annotations.AfterClass;
import org.testng.annotations.BeforeClass;
import org.testng.annotations.Test;

public class SaveMobilePhoneTest {
```

```
private CloseableHttpClient client;
private CloseableHttpResponse response;

@BeforeClass
public void init() {
    client = HttpClients.createDefault();
}

@Test
public void testCase1() {
    Assert.assertEquals("{\"code\":0,\"message\":\"保存成功！\"}",
        sendHttpPostRequest(client, "{\"brand\":\"Motorola\",\"model\":\"
           moto Z Play\",\"os\":\"ANDROID\"}"));
}

@Test
public void testCase2() {
    Assert.assertEquals("{\"code\":1,\"message\":\"保存失败！\"}",
        sendHttpPostRequest(client, "{\"brand\":\"\",\"model\":\"moto Z
           Play\",\"os\":\"ANDROID\"}"));
}

@Test
public void testCase3() {
    Assert.assertEquals("{\"code\":1,\"message\":\"保存失败！\"}",
        sendHttpPostRequest(client, "{\"brand\":null,\"model\":\"moto Z
           Play\",\"os\":\"ANDROID\"}"));
}

@Test
public void testCase4() {
    Assert.assertEquals("{\"code\":1,\"message\":\"保存失败！\"}",
        sendHttpPostRequest(client, "{\"brand\":\"Motorola\",\"
           model\":\"\",\"os\":\"ANDROID\"}"));
}

@Test
public void testCase5() {
    Assert.assertEquals("{\"code\":1,\"message\":\"保存失败！\"}",
        sendHttpPostRequest(client, "{\"brand\":\"Motorola\",\"
           model\":null,\"os\":\"ANDROID\"}"));
}

@Test
public void testCase6() {
    Assert.assertEquals("{\"code\":1,\"message\":\"保存失败！\"}",
        sendHttpPostRequest(client, "{\"brand\":\"Motorola\",\"model\":\"
           moto Z Play\",\"os\":\"\"}"));
}
```

```
@Test
public void testCase7() {
    Assert.assertEquals("{\"code\":1,\"message\":\"保存失败! \"}",
        sendHttpPostRequest(client, "{\"brand\":\"Motorola\",\"model\":\"
            moto Z Play\",\"os\":null}"));
}

@Test
public void testCase8() {
    Assert.assertEquals("{\"code\":0,\"message\":\"保存成功! \"}",
       sendHttpPostRequest(client,
            "{\"brand\":\"
                0123456789012345678901234567890123456789012345678 9\
                ",\"model\":\"moto Z Play\",\"os\":\"ANDROID\"}"));
}

@Test
public void testCase9() {
    Assert.assertEquals("{\"code\":0,\"message\":\"保存成功! \"}",
        sendHttpPostRequest(client,
        "{\"brand\":\"Motorola\",\"model\":\"
            01234567890123456789012345678901234567890123456789\
            ",\"os\":\"ANDROID\"}"));
}

@Test
public void testCase10() {
    Assert.assertEquals("{\"code\":0,\"message\":\"保存成功! \"}",
        sendHttpPostRequest(client, "{\"brand\":\"Apple\",\"
            model\":\"iPhone XS Max\",\"os\":\"IOS\"}"));
}

@Test
public void testCase11() {
    Assert.assertEquals("{\"code\":1,\"message\":\"保存失败! \"}",
        sendHttpPostRequest(client, "{\"brand\":\"Nokia\",\"model\":\"
            N95\",\"os\":\"SYMBIAN\"}"));
}

@AfterClass
public void clear() {
    try {
        response.close();
        client.close();
    } catch (Exception e) {
        e.printStackTrace();
    }
}
```

```
    private String sendHttpPostRequest(CloseableHttpClient client, String
        stringEntity) {
        String result = null;
        try {
            HttpPost httpPost = new
                HttpPost("http://localhost:8080/mobilePhone");
            httpPost.setEntity(new StringEntity(stringEntity,
                ContentType.APPLICATION_JSON));
            response = client.execute(httpPost);
            result = EntityUtils.toString(response.getEntity());
        } catch (Exception e) {
            e.printStackTrace();
        }
        return result;
    }

}
```

修改 testng.xml 文件，修改<class>标签中的 name 属性值，见以下粗体部分内容：

```
<?xml version="1.0" encoding="UTF-8"?>
<!DOCTYPE suite SYSTEM "http://testng.org/testng-1.0.dtd">
<suite name="Suite">
    <test thread-count="5" name="Test">
        <classes>
            <class
                name="com.lujiatao.httpinterfacetest.SaveMobilePhoneTest" />
        </classes>
    </test> <!-- Test -->
</suite> <!-- Suite -->
```

保存所做的修改，在 testng.xml 上用鼠标右击，从弹出的快捷菜单中选择“Run As → TestNG Suite”选项，然后查看测试报告，如图 4-3 所示。

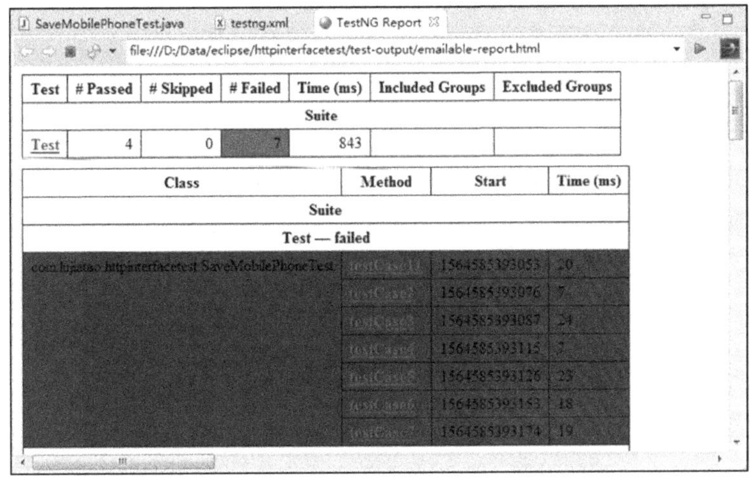

Test	# Passed	# Skipped	# Failed	Time (ms)	Included Groups	Excluded Groups
Suite						
Test	4	0	7	843		

Class	Method	Start	Time (ms)
Suite			
Test — failed			
com.lujiatao.httpinterfacetest.SaveMobilePhoneTest	[illegible]	1564585393053	20
	[illegible]	1564585393076	7
	[illegible]	1564585393087	24
	[illegible]	1564585393115	7
	[illegible]	1564585393126	23
	[illegible]	1564585393163	18
	[illegible]	1564585393174	19

图 4-3

下面对运行结果进行说明。

① 整体思路和上一个示例一致，只是将 sendHttpGetRequest() 方法改成了 sendHttpPostRequest()方法，具体的实现都封装到了 sendHttpPostRequest()方法中。

② 从测试报告可以看出，有 7 个测试用例执行失败了，这时我们可以查看失败原因，以 testCase11 为例，单击“testCase11”，如图 4-4 所示。

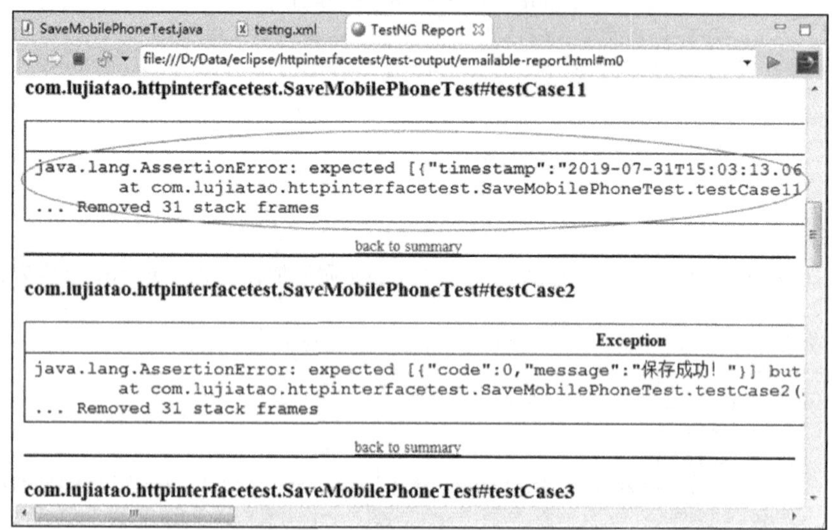

图 4-4

复制 expected 后的方括号中的内容：

```
{
    "timestamp": "2019-07-31T15:03:13.069+0000",
    "status": 500,
    "error": "Internal Server Error",
    "message": "No enum constant com.lujiatao.httpinterface.domain.Os.SYMBIAN",
    "path": "/mobilePhone"
}
```

可以看到服务器返回了 500 错误，提示没有“SYMBIAN”这个枚举，这里表明了待测程序设计不够严谨。待测程序没有对入参进行校验，导致不能返回友好的提示，而是直接返回了服务器 500 错误。

实际项目中，自动化测试用例执行失败后，通过对结果进行分析，在确认失败原因为待测程序的缺陷时，就应该将缺陷提交给开发人员。

3．简化自动化测试用例

上例中有 11 个自动化测试用例，但通过观察可以发现入参和期望值很相似，因此代码看上去很冗余，有没有简化的方法呢？答案是肯定的。第 3 章曾介绍过数据分离，这里就派上用场了。删除 SaveMobilePhoneTest 中的内容，输入以下内容：

```
package com.lujiatao.httpinterfacetest;

import org.apache.http.client.methods.CloseableHttpResponse;
import org.apache.http.client.methods.HttpPost;
import org.apache.http.entity.ContentType;
import org.apache.http.entity.StringEntity;
import org.apache.http.impl.client.CloseableHttpClient;
import org.apache.http.impl.client.HttpClients;
import org.apache.http.util.EntityUtils;
import org.testng.Assert;
import org.testng.annotations.AfterClass;
import org.testng.annotations.BeforeClass;
import org.testng.annotations.DataProvider;
import org.testng.annotations.Test;

public class SaveMobilePhoneTest {
    private CloseableHttpClient client;
    private CloseableHttpResponse response;

    @BeforeClass
    public void init() {
        client = HttpClients.createDefault();
    }

    @DataProvider(name = "data")
    public Object[][] dp() {
        Object[][] data = new Object[11][];
        String success = "{\"code\":0,\"message\":\"保存成功！\"}";
        String fail = "{\"code\":1,\"message\":\"保存失败！\"}";
        data[0] = new Object[] { success,
           "{\"brand\":\"Motorola\",\"model\":\"moto Z
           Play\",\"os\":\"ANDROID\"}" };
        data[1] = new Object[] { fail, "{\"brand\":\"\",\"model\":\"moto Z
           Play\",\"os\":\"ANDROID\"}" };
        data[2] = new Object[] { fail, "{\"brand\":null,\"model\":\"moto Z
           Play\",\"os\":\"ANDROID\"}" };
        data[3] = new Object[] { fail,
           "{\"brand\":\"Motorola\",\"model\":\"\",\"os\":\"ANDROID\"}" };
        data[4] = new Object[] { fail,
           "{\"brand\":\"Motorola\",\"model\":null,\"os\":\"ANDROID\"}" };
        data[5] = new Object[] { fail, "{\"brand\":\"Motorola\",\"model\":\"moto
           Z Play\",\"os\":\"\"}" };
        data[6] = new Object[] { fail, "{\"brand\":\"Motorola\",\"model\":\"moto
           Z Play\",\"os\":null}" };
        data[7] = new Object[] { success,
             "{\"brand\":\" 0123456789012345678901234567890123456789012345678 9\
             ",\"model\":\"moto Z Play\",\"os\":\"ANDROID\"}" };
        data[8] = new Object[] { success,
```

```
                "{\"brand\":\"Motorola\",\"model\":\"
                01234567890123456789012345678901234567890123456789\
                ",\"os\":\"ANDROID\"}" };
        data[9] = new Object[] { success, "{\"brand\":\"Apple\",\"model\":\"iPhone
            XS Max\",\"os\":\"IOS\"}" };
        data[10] = new Object[] { fail,
            "{\"brand\":\"Nokia\",\"model\":\"N95\",\"os\":\"SYMBIAN\"}" };
        return data;
    }

    @Test(dataProvider = "data")
    public void testCase1(String expected, String stringEntity) {
        Assert.assertEquals(expected, sendHttpPostRequest(client, stringEntity));
    }

    @AfterClass
    public void clear() {
        try {
            response.close();
            client.close();
        } catch (Exception e) {
            e.printStackTrace();
        }
    }

    private String sendHttpPostRequest(CloseableHttpClient client, String stringEntity) {
        String result = null;
        try {
            HttpPost httpPost = new
                HttpPost("http://localhost:8080/mobilePhone");
            httpPost.setEntity(new StringEntity(stringEntity,
                ContentType.APPLICATION_JSON));
            response = client.execute(httpPost);
            result = EntityUtils.toString(response.getEntity());
        } catch (Exception e) {
            e.printStackTrace();
        }
        return result;
    }

}
```

保存代码，选择“Run As → TestNG Suite”选项，测试报告如图 4-5 所示。

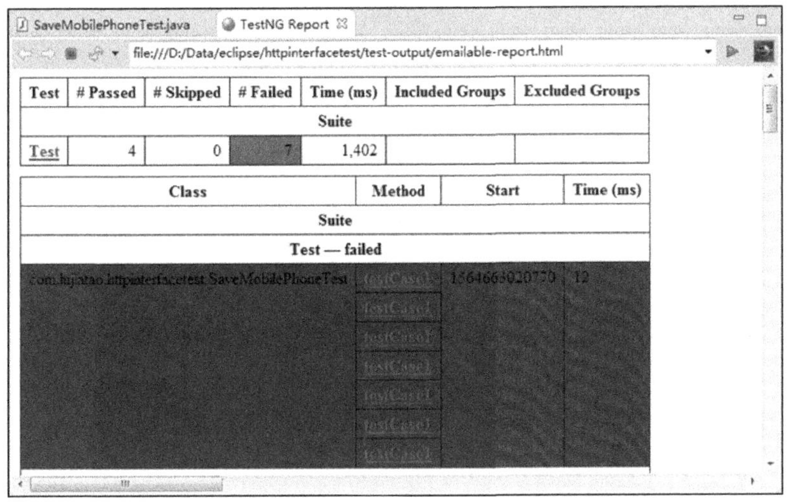

图 4-5

这里使用了@DataProvider 注解修饰一个方法，该方法用于提供测试数据，测试数据包括期望值和请求的实体字符串。

4. 解析 JSON 字符串

RESTful 接口通过 JSON 传输数据，以上的例子我们都将 JSON 字符串当作“一般字符串”对待了，实际上可以通过对 JSON 字符串的解析进行更细粒度的控制。作为示例讲解，这里删除了 testCase2～testCase11，仅保留 testCase1：

```
package com.lujiatao.httpinterfacetest;

import org.apache.http.client.methods.CloseableHttpResponse;
import org.apache.http.client.methods.HttpPost;
import org.apache.http.entity.ContentType;
import org.apache.http.entity.StringEntity;
import org.apache.http.impl.client.CloseableHttpClient;
import org.apache.http.impl.client.HttpClients;
import org.apache.http.util.EntityUtils;
import org.json.JSONObject;
import org.testng.Assert;
import org.testng.annotations.AfterClass;
import org.testng.annotations.BeforeClass;
import org.testng.annotations.Test;

public class SaveMobilePhoneTest {
    private CloseableHttpClient client;
    private CloseableHttpResponse response;

    @BeforeClass
    public void init() {
```

```
        client = HttpClients.createDefault();
    }

    @Test
    public void testCase1() {
        JSONObject expected = new JSONObject().put("code", 0).put("message",
           "保存成功！");
        JSONObject stringEntity = new JSONObject().put("brand",
           "Motorola").put("model", "moto Z Play").put("os", "ANDROID");
        JSONObject actual = sendHttpPostRequest(client, stringEntity.toString());
        if (!(expected.get("code") == actual.get("code"))
           || !(expected.get("message").equals(actual.get("message")))) {
            Assert.fail("失败！");
        }
    }

    @AfterClass
    public void clear() {
        try {
            response.close();
            client.close();
        } catch (Exception e) {
            e.printStackTrace();
        }
    }

    private JSONObject sendHttpPostRequest(CloseableHttpClient client, String
        stringEntity) {
        JSONObject result = null;
        try {
            HttpPost httpPost = new HttpPost("http://localhost:8080/mobilePhone");
            httpPost.setEntity(new StringEntity(stringEntity,
                ContentType.APPLICATION_JSON));
            response = client.execute(httpPost);
            result = new JSONObject(EntityUtils.toString(response.getEntity()));
        } catch (Exception e) {
            e.printStackTrace();
        }
        return result;
    }

}
```

在 testCase1 中，将预期结果、入参和实际结果都用 JSONObject 对象表示，包括断言也是通过操作 JSONObject 对象完成的，整个过程看不到“糟心”的字符串。这样做的好处是将 RESTful 请求的请求体和响应体完全“JSON”化，便于进一步封装，并且与字符串也实现了“解耦”。

4.5.2　SOAP 接口自动化测试

1．自动化测试用例编写

在 com.lujiatao.httpinterfacetest Package 中创建名为 GetMobilePhoneSoapTest 的 Class，在 GetMobilePhoneSoapTest 中输入以下内容：

```
package com.lujiatao.httpinterfacetest;

import org.apache.http.client.methods.CloseableHttpResponse;
import org.apache.http.client.methods.HttpPost;
import org.apache.http.entity.ContentType;
import org.apache.http.entity.StringEntity;
import org.apache.http.impl.client.CloseableHttpClient;
import org.apache.http.impl.client.HttpClients;
import org.apache.http.util.EntityUtils;
import org.testng.Assert;
import org.testng.annotations.AfterClass;
import org.testng.annotations.BeforeClass;
import org.testng.annotations.Test;

public class GetMobilePhoneSoapTest {
    private CloseableHttpClient client;
    private CloseableHttpResponse response;

    @BeforeClass
    public void init() {
        client = HttpClients.createDefault();
    }

    @Test
    public void testCase1() {
        String expected = "<SOAP-ENV:Envelope
           xmlns:SOAP-ENV=\"http://schemas.xmlsoap……
           \"><SOAP-ENV:Header/><SOAP-ENV:Body><ns2:getMobilePhoneResponse
           xmlns:ns2=\"http://www.lujiatao……
           \"><ns2:mobilePhone><ns2:brand>Apple</ns2:brand><ns2:model>iPhone
           6S</ns2:model><ns2:os>IOS</ns2:os></ns2:mobilePhone></ns2:getMobile
           PhoneResponse></SOAP-ENV:Body></SOAP-ENV:Envelope>";
        String soapString = "<SOAP-ENV:Envelope
           xmlns:SOAP-ENV=\"http://schemas.xmlsoap.org/soap/envelope/\"
           xmlns:hi=\"http://www.lujiatao……\">\r\n" + "
           <SOAP-ENV:Header/>\r\n" + "  <SOAP-ENV:Body>\r\n" + "
           <hi:getMobilePhoneRequest>\r\n" + " <hi:model>iPhone
           6S</hi:model>\r\n" + " </hi:getMobilePhoneRequest>\r\n" + "
           </SOAP-ENV:Body>\r\n" + "</SOAP-ENV:Envelope>";
        Assert.assertEquals(expected, sendHttpPostRequest(client, soapString));
    }

    @Test
    public void testCase2() {
```

```
        String expected = "<SOAP-ENV:Envelope
            xmlns:SOAP-ENV=\"http://schemas.xmlsoap.org/soap/envelope/\"><SOAP-
            ENV:Header/><SOAP-ENV:Body><ns2:getMobilePhoneResponse
            xmlns:ns2=\"http://www.lujiatao.com/httpinterface/MobilePhones\"/><
            /SOAP-ENV:Body></SOAP-ENV:Envelope>";
        String soapString = "<SOAP-ENV:Envelope
            xmlns:SOAP-ENV=\"http://schemas.xmlsoap.org/soap/envelope/\"
            xmlns:hi=\"http://www.lujiatao.com/httpinterface/MobilePhones\">\r\
            n" + "    <SOAP-ENV:Header/>\r\n" + "    <SOAP-ENV:Body>\r\n" + "
            <hi:getMobilePhoneRequest>\r\n" + " <hi:model></hi:model>\r\n" + "
            </hi:getMobilePhoneRequest>\r\n" + "    </SOAP-ENV:Body>\r\n" +
            "</SOAP-ENV:Envelope>";
        Assert.assertEquals(expected, sendHttpPostRequest(client, soapString));
    }

    @Test
    public void testCase3() {
        String expected = "<SOAP-ENV:Envelope
            xmlns:SOAP-ENV=\"http://schemas.xmlsoap.org/soap/envelope/\"><SOAP-
            ENV:Header/><SOAP-ENV:Body><ns2:getMobilePhoneResponse
            xmlns:ns2=\"http://www.lujiatao.com/httpinterface/MobilePhones\"/><
            /SOAP-ENV:Body></SOAP-ENV:Envelope>";
        String soapString = "<SOAP-ENV:Envelope
            xmlns:SOAP-ENV=\"http://schemas.xmlsoap.org/soap/envelope/\"
            xmlns:hi=\"http://www.lujiatao.com/httpinterface/MobilePhones\">\r\
            n" + "    <SOAP-ENV:Header/>\r\n" + "    <SOAP-ENV:Body>\r\n" + "
            <hi:getMobilePhoneRequest>\r\n" + "<hi:model>null</hi:model>\r\n" + "
            </hi:getMobilePhoneRequest>\r\n" + "    </SOAP-ENV:Body>\r\n" +
            "</SOAP-ENV:Envelope>";
        Assert.assertEquals(expected, sendHttpPostRequest(client, soapString));
    }

    @Test
    public void testCase4() {
        String expected = "<SOAP-ENV:Envelope
            xmlns:SOAP-ENV=\"http://schemas.xmlsoap……
            "><SOAP-ENV:Header/><SOAP-ENV:Body><ns2:getMobilePhoneResponse
            xmlns:ns2=\"http://www.lujiatao……
            "/></SOAP-ENV:Body></SOAP-ENV:Envelope>";
        String soapString = "<SOAP-ENV:Envelope
            xmlns:SOAP-ENV=\"http://schemas.xmlsoap……"
            xmlns:hi=\"http://www.lujiatao……">\r\n" + "
            <SOAP-ENV:Header/>\r\n" + "    <SOAP-ENV:Body>\r\n" + "
            <hi:getMobilePhoneRequest>\r\n" + "
            <hi:model>01234567890123456789012345678901234567890123456789</hi:mo
            del>\r\n" + "        </hi:getMobilePhoneRequest>\r\n" + "
            </SOAP-ENV:Body>\r\n" + "</SOAP-ENV:Envelope>";
        Assert.assertEquals(expected, sendHttpPostRequest(client, soapString));
    }

    @AfterClass
    public void clear() {
        try {
```

```
            response.close();
            client.close();
        } catch (Exception e) {
            e.printStackTrace();
        }
    }

    private String sendHttpPostRequest(CloseableHttpClient client, String
        soapString) {
        String result = null;
        try {
            HttpPost httpPost = new
                HttpPost("http://localhost:8080/MobilePhones");
            httpPost.setEntity(new StringEntity(soapString,
                ContentType.TEXT_XML));
            response = client.execute(httpPost);
            result = EntityUtils.toString(response.getEntity());
        } catch (Exception e) {
            e.printStackTrace();
        }
        return result;
    }

}
```

修改 testng.xml 文件，修改<class>标签中的 name 属性值，见以下粗体部分内容：

```
<?xml version="1.0" encoding="UTF-8"?>
<!DOCTYPE suite SYSTEM "http://testng.org/testng-1.0.dtd">
<suite name="Suite">
    <test thread-count="5" name="Test">
        <classes>
            <class
                name="com.lujiatao.httpinterfacetest.GetMobilePhoneSoapTest" />
        </classes>
    </test> <!-- Test -->
</suite> <!-- Suite -->
```

保存所做的修改，在 testng.xml 上用鼠标右击，从弹出的快捷菜单中选择“Run As → TestNG Suite”选项，测试报告如图 4-6 所示。

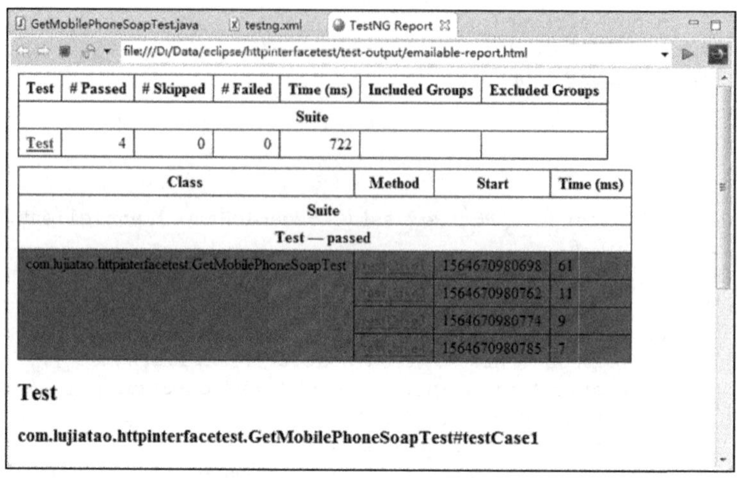

Test	# Passed	# Skipped	# Failed	Time (ms)	Included Groups	Excluded Groups
Suite						
Test	4	0	0	722		

Class	Method	Start	Time (ms)
Suite			
Test — passed			
com.lujiatao.httpinterfacetest.GetMobilePhoneSoapTest	[illegible]	1564670980698	61
	[illegible]	1564670980762	11
	[illegible]	1564670980774	9
	[illegible]	1564670980785	7

Test

com.lujiatao.httpinterfacetest.GetMobilePhoneSoapTest#testCase1

图 4-6

测试用例均通过测试，与预期一致。

2. 解析 XML 字符串

直接处理 XML 字符串虽然能解决问题，但是代码可读性很差。这里使用 DOM4J 处理 XML 字符串。作为示例讲解，这里仅保留了 testCase1：

```
package com.lujiatao.httpinterfacetest;

import java.util.HashMap;
import java.util.Map;

import org.apache.http.client.methods.CloseableHttpResponse;
import org.apache.http.client.methods.HttpPost;
import org.apache.http.entity.ContentType;
import org.apache.http.entity.StringEntity;
import org.apache.http.impl.client.CloseableHttpClient;
import org.apache.http.impl.client.HttpClients;
import org.apache.http.util.EntityUtils;
import org.dom4j.Document;
import org.dom4j.DocumentHelper;
import org.dom4j.Element;
import org.dom4j.XPath;
import org.testng.Assert;
import org.testng.annotations.AfterClass;
import org.testng.annotations.BeforeClass;
import org.testng.annotations.Test;

public class GetMobilePhoneSoapTest {
    private CloseableHttpClient client;
    private CloseableHttpResponse response;
```

```java
@BeforeClass
public void init() {
    client = HttpClients.createDefault();
}

@Test
public void testCase1() {
    // 构建期望 XML
    Document expected = DocumentHelper.createDocument();
    Element root = expected.addElement("SOAP-ENV:Envelope").
        addAttribute("xmlns:SOAP-ENV",
            "http://schemas.xmlsoap:……");
    root.addNamespace("SOAP-ENV", "http://schemas.xmlsoap……");
    root.addElement("SOAP-ENV:Header");
    Element body = root.addElement("SOAP-ENV:Body");
    Element node = body.addNamespace("ns2", "http://www.lujiatao……")
            .addElement("ns2:getMobilePhoneResponse")
            .addAttribute("xmlns:ns2", "http://www.lujiatao……");
    Element nodeChild = node.addElement("ns2:mobilePhone");
    nodeChild.addElement("ns2:brand").addText("Apple");
    nodeChild.addElement("ns2:model").addText("iPhone 6S");
    nodeChild.addElement("ns2:os").addText("IOS");
    // 构建期望 XML 对应的 XPath
    Map<String, String> xmlMap = new HashMap<>();
    xmlMap.put("ns2", "http://www.lujiatao……");
    XPath xPath1 = expected.createXPath("//ns2:brand");
    XPath xPath2 = expected.createXPath("//ns2:model");
    XPath xPath3 = expected.createXPath("//ns2:os");
    xPath1.setNamespaceURIs(xmlMap);
    xPath2.setNamespaceURIs(xmlMap);
    xPath3.setNamespaceURIs(xmlMap);
    // 构建入参 XML
    Document soapString = DocumentHelper.createDocument();
    Element root2 = soapString.addElement("SOAP-ENV:Envelope")
            .addAttribute("xmlns:SOAP-ENV", "http://schemas.xmlsoap……")
            .addAttribute("xmlns:hi", "http://www.lujiatao……");
    root2.addNamespace("SOAP-ENV", "http://schemas.xmlsoap……");
    root2.addNamespace("hi", "http://www.lujiatao……");
    root2.addElement("SOAP-ENV:Header");
    Element body2 = root2.addElement("SOAP-ENV:Body");
    Element node2 = body2.addElement("hi:getMobilePhoneRequest");
    node2.addElement("hi:model").addText("iPhone 6S");
    // 接收响应 XML 并断言
    Document actual = sendHttpPostRequest(client, soapString.asXML());
    if (!(xPath1.selectSingleNode(expected).getText().equals(xPath1.
       selectSingleNode(actual).getText()))
            || !(xPath2.selectSingleNode(expected).getText().equals(xPath2.
               selectSingleNode(actual).getText()))
```

```
                || !(xPath3.selectSingleNode(expected).getText().
                    equals(xPath3.selectSingleNode(actual).getText()))) {
                Assert.fail("失败! ");
            }
        }

        @AfterClass
        public void clear() {
            try {
                response.close();
                client.close();
            } catch (Exception e) {
                e.printStackTrace();
            }
        }

        private Document sendHttpPostRequest(CloseableHttpClient client, String
            soapString) {
            Document result = null;
            try {
                HttpPost httpPost = new
                    HttpPost("http://localhost:8080/MobilePhones");
                httpPost.setEntity(new StringEntity(soapString,
                    ContentType.TEXT_XML));
                response = client.execute(httpPost);
                result = DocumentHelper.parseText(EntityUtils.
                    toString(response.getEntity()));
            } catch (Exception e) {
                e.printStackTrace();
            }
            return result;
        }

    }
```

下面对代码进行说明。

① 使用 DOM4J 中的 API 构建期望值，实际上一个是 Document 对象。

② 后续断言会使用 XPath 方式查找期望值和实际值中的指定数据，因此这里先构建 XPath。

③ 使用 DOM4J 中的 API 构建入参。

④ 首先接收实际值（也是一个 Document 对象），然后通过 XPath 获得节点，最后通过对节点的文本进行比较完成断言。

第 5 章 RPC 接口自动化测试

5.1 RPC 简介

RPC（Remote Procedure Call，远程过程调用）协议是一种计算机通信协议，该协议允许本地计算机上的程序远程调用另一台计算机上的程序，而在此过程中，开发人员不需要额外编程。另外，在面向对象编程的程序中，远程过程调用又可称为远程调用或远程方法调用。常见的 RPC 协议实现如下。

Java RMI：Java Remote Method Invocation，即 Java 远程方法调用，采用了 JDK 中标准的 java.rmi.*实现。

Apache Dubbo（以下简称 Dubbo）：是阿里巴巴开源的一款高性能、轻量级的 Java RPC 框架。Dubbo 的特点包括可面向接口的远程方法调用、智能容错、负载均衡和服务自动注册及发现等。

Hessian：Hessian 是一种二进制协议，它非常适合发送二进制数据，且无须通过附件的形式对该协议进行任何扩展。

XML-RPC：XML-RPC 是一个规范和一组实现，它允许程序运行在不同的操作系统和环境中，以通过 Internet 进行过程调用。它是一种基于 HTTP 和 XML 实现的远程过程调用。XML-RPC 的设计虽然非常简单，但它支持传输、处理和返回复杂的数据结构。

Apache Thrift：是一种 RPC 协议实现的框架，适用于跨语言服务的开发，它将软件堆栈与代码生成引擎结合起来构建高效的服务，且可无缝连接 Java、Python、C++、JavaScript、C#和 PHP 等编程语言。

在众多 RPC 协议的实现中，Dubbo 作为分布式微服务的主流框架之一，在企业级应用开发中使用频率较高，因此本章使用 Dubbo 作为 RPC 接口自动化测试的示例。

Dubbo 中的重要角色如图 5-1 所示。

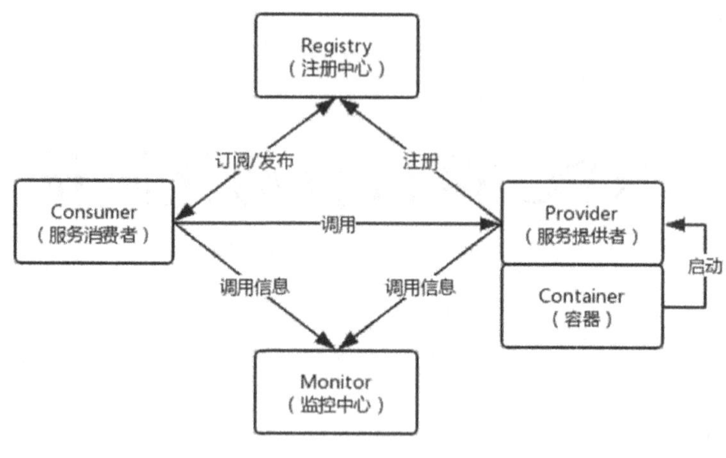

图 5-1

服务提供者：提供远程调用的接口。

容器：运行服务提供者。

服务消费者：调用服务提供者提供的接口。

注册中心：服务提供者将服务注册到注册中心，服务消费者从注册中心获取需要使用的服务。

监控中心：监控服务提供者和服务消费者的调用过程。监控中心是可选的。

5.2 部署待测程序

1. 安装 JDK

在即将部署待测程序的服务器上安装 JDK，并配置系统变量如下。这里以 Windows 为例。

① 新建 JAVA_HOME，变量值填写如下。

```
C:\Program Files\Java\jdk1.8.0_212
```

② 新建 CLASSPATH，变量值填写如下。

```
.;%JAVA_HOME%\lib\dt.jar;%JAVA_HOME%\lib\tools.jar
```

③ 编辑 Path，变量值追加如下。

```
;%JAVA_HOME%\bin
```

2. 安装 ZooKeeper

在 Dubbo 中，注册中心只是一个角色，可以充当该角色的中间件有很多，ZooKeeper 就是其中最常用的一个，同时也是官方推荐使用的。ZooKeeper 可以部署为单机模式或分布式模式，为简明步骤，示例部署为单机模式。这里以 Windows 为例。

① 从官网下载 ZooKeeper。

② 解压缩到指定目录，笔者解压缩到 D:\Program Files 目录。

③ 将 conf 目录下的 zoo_sample.cfg 文件重命名为 zoo.cfg，编辑 zoo.cfg 文件，在文件最后加上 admin.serverPort=8081。

④ 双击 bin 目录下的 zkServer.cmd 文件，运行成功后如图 5-2 所示。

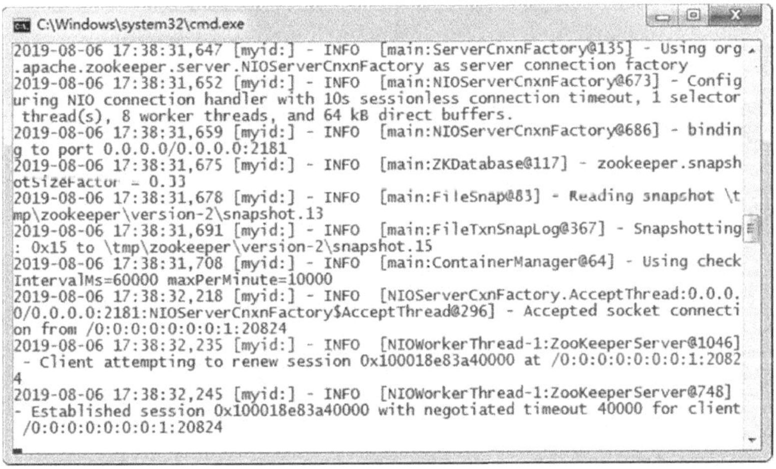

```
C:\Windows\system32\cmd.exe
2019-08-06 17:38:31,647 [myid:] - INFO  [main:ServerCnxnFactory@135] - Using org
.apache.zookeeper.server.NIOServerCnxnFactory as server connection factory
2019-08-06 17:38:31,652 [myid:] - INFO  [main:NIOServerCnxnFactory@673] - Config
uring NIO connection handler with 10s sessionless connection timeout, 1 selector
 thread(s), 8 worker threads, and 64 kB direct buffers.
2019-08-06 17:38:31,659 [myid:] - INFO  [main:NIOServerCnxnFactory@686] - bindin
g to port 0.0.0.0/0.0.0.0:2181
2019-08-06 17:38:31,675 [myid:] - INFO  [main:ZKDatabase@117] - zookeeper.snapsh
otSizeFactor = 0.33
2019-08-06 17:38:31,678 [myid:] - INFO  [main:FileSnap@83] - Reading snapshot \t
mp\zookeeper\version-2\snapshot.13
2019-08-06 17:38:31,691 [myid:] - INFO  [main:FileTxnSnapLog@367] - Snapshotting
: 0x15 to \tmp\zookeeper\version-2\snapshot.15
2019-08-06 17:38:31,708 [myid:] - INFO  [main:ContainerManager@64] - Using check
IntervalMs=60000 maxPerMinute=10000
2019-08-06 17:38:32,218 [myid:] - INFO  [NIOServerCxnFactory.AcceptThread:0.0.0.
0/0.0.0.0:2181:NIOServerCnxnFactory$AcceptThread@296] - Accepted socket connecti
on from /0:0:0:0:0:0:0:1:20824
2019-08-06 17:38:32,235 [myid:] - INFO  [NIOWorkerThread-1:ZooKeeperServer@1046]
 - Client attempting to renew session 0x100018e83a40000 at /0:0:0:0:0:0:0:1:2082
4
2019-08-06 17:38:32,245 [myid:] - INFO  [NIOWorkerThread-1:ZooKeeperServer@748]
- Established session 0x100018e83a40000 with negotiated timeout 40000 for client
 /0:0:0:0:0:0:0:1:20824
```

图 5-2

3. 部署待测程序

从笔者的 GitHub 中下载待测程序，程序名为 rpcinterface-0.0.1-SNAPSHOT.jar，下载后放在服务器上，执行以下命令运行即可。

```
java -jar E:\rpcinterface-0.0.1-SNAPSHOT.jar
```

这里把本地电脑作为服务器，且待测程序放在了 E 盘根目录，读者需要根据实际情况替换以上路径。运行成功后如图 5-3 所示。

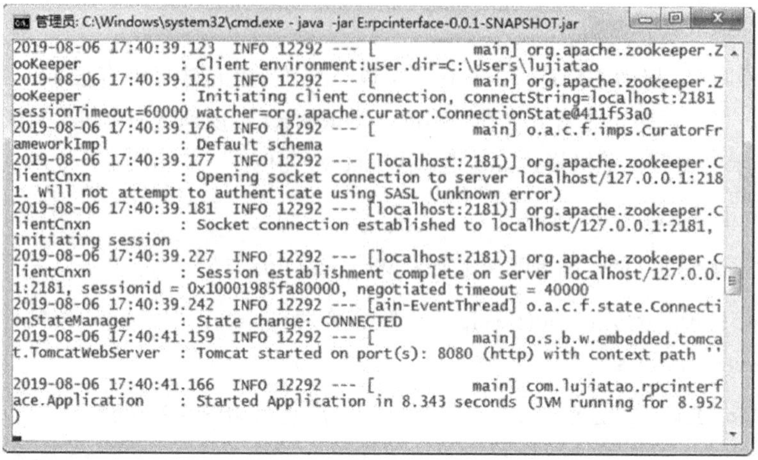

```
管理员: C:\Windows\system32\cmd.exe - java -jar E:rpcinterface-0.0.1-SNAPSHOT.jar
2019-08-06 17:40:39.123  INFO 12292 --- [           main] org.apache.zookeeper.Z
ooKeeper             : Client environment:user.dir=C:\Users\lujiatao
2019-08-06 17:40:39.125  INFO 12292 --- [           main] org.apache.zookeeper.Z
ooKeeper             : Initiating client connection, connectString=localhost:2181
sessionTimeout=60000 watcher=org.apache.curator.ConnectionState@411f53a0
2019-08-06 17:40:39.176  INFO 12292 --- [           main] o.a.c.f.imps.CuratorFr
ameworkImpl          : Default schema
2019-08-06 17:40:39.177  INFO 12292 --- [localhost:2181)] org.apache.zookeeper.C
lientCnxn            : Opening socket connection to server localhost/127.0.0.1:218
1. Will not attempt to authenticate using SASL (unknown error)
2019-08-06 17:40:39.181  INFO 12292 --- [localhost:2181)] org.apache.zookeeper.C
lientCnxn            : Socket connection established to localhost/127.0.0.1:2181,
initiating session
2019-08-06 17:40:39.227  INFO 12292 --- [localhost:2181)] org.apache.zookeeper.C
lientCnxn            : Session establishment complete on server localhost/127.0.0.
1:2181, sessionid = 0x10001985fa80000, negotiated timeout = 40000
2019-08-06 17:40:39.242  INFO 12292 --- [ain-EventThread] o.a.c.f.state.Connecti
onStateManager       : State change: CONNECTED
2019-08-06 17:40:41.159  INFO 12292 --- [           main] o.s.b.w.embedded.tomca
t.TomcatWebServer    : Tomcat started on port(s): 8080 (http) with context path ''

2019-08-06 17:40:41.166  INFO 12292 --- [           main] com.lujiatao.rpcinterf
ace.Application      : Started Application in 8.343 seconds (JVM running for 8.952
)
```

图 5-3

5.3　手工测试用例设计

5.3.1　分析待测接口

Dubbo 接口的分析方式和 HTTP 接口类似，接口文档如表 5-1 所示。

表 5-1

1. 通过手机型号获取手机	
接口路径	com.lujiatao.rpcinterface.dubbo.MobilePhoneService
接口方法	getMobilePhone
参数	String model
返回值	com.lujiatao.rpcinterface.domain.MobilePhone
返回值示例	{"os":"IOS","model":"iPhone 6S","brand":"Apple"}
2. 保存手机	
接口路径	com.lujiatao.rpcinterface.dubbo.MobilePhoneService
接口方法	saveMobilePhone
参数	com.lujiatao.rpcinterface.domain.MobilePhone mobilePhone
返回值	String
返回值示例	{"code":0,"message":"保存成功！"}

从接口中可以看出待测程序只有一个接口，但有两个方法，即 getMobilePhone 和 saveMobilePhone，这与第 4 章中的待测程序相同。其中，com.lujiatao.rpcinterface.domain.MobilePhone 为一个实体类，由于不是 Java 的内置类型，因此在接口文档中最好写出全路径。

5.3.2　测试用例设计

这里仅以入参作为示例来设计测试用例，5.4 节会将这部分手工测试用例转换成自动化测试用例。

1．getMobilePhone 测试用例设计

（1）参数必填项

Case 1：model="iPhone 6S"——返回 iPhone 6S 手机（假设存在 iPhone 6S）。

Case 2：model=""——返回空。

Case 3：model=null——返回空。

（2）参数长度

接口文档未提及参数长度，但我们可以提交一个相对较长的参数。

Case 4：model="01234567890123456789012345678901234567890123456789"——返回空。

（3）参数组合

在多个参数接口中，参数可能相互响应或制约，因此对参数组合的用例设计必不可少。本接口不涉及（因为只有一个参数）。

（4）参数规则

一般特殊的参数会有相应规则，比如手机号、身份证号和统一社会信用代码等。本接口不涉及。

（5）参数枚举

有些参数只能接收固定的几个参数值，这类参数在服务端大多用枚举定义，在这种情况下，需要测试每种枚举值。本接口不涉及。

综合以上各种情况，getMobilePhone 一共设计了 4 条手工测试用例。

2．saveMobilePhone 测试用例设计

（1）参数必填项

Case 1：{"brand":"Motorola","model":"moto Z Play","os":"ANDROID"}——保存成功。

Case 2：{"brand":"","model":"moto Z Play","os":"ANDROID"}——保存失败。

Case 3：{"brand":null,"model":"moto Z Play","os":"ANDROID"}——保存失败。

Case 4：{"brand":"Motorola","model":"","os":"ANDROID"}——保存失败。

Case 5：{"brand":"Motorola","model":null,"os":"ANDROID"}——保存失败。

Case 6：{"brand":"Motorola","model":"moto Z Play","os":""}——保存失败。

Case 7：{"brand":"Motorola","model":"moto Z Play","os":null}——保存失败。

（2）参数长度

接口文档未提及参数长度，但我们可以提交一个相对较长的参数。

Case 8：

{"brand":"01234567890123456789012345678901234567890123456789","model":"moto Z Play","os":"ANDROID"}——保存成功。

Case 9：

{"brand":"Motorola","model":"01234567890123456789012345678901234567890123456789","os":"ANDROID"}——保存成功。

这里没有对 os 参数做参数长度测试，稍后会说明原因。

（3）参数组合

本接口不涉及。

（4）参数规则

本接口不涉及。

（5）参数枚举

os 参数代表操作系统，而手机操作系统目前主流的只有 Android 和 iOS，且在接口文档的示例中，os 的值为“IOS”。在 Java 中，枚举用大写表示，故推测该字段在后台以枚举方式定义。

Case 10：{"brand":"Apple","model":"iPhone XS Max","os":"IOS"}——保存成功。

Case 11：{"brand":"Nokia","model":"N95","os":"SYMBIAN"}——保存失败。

os 的值为“ANDROID”，上述已有用例覆盖，因此这里不再单独设计用例进行覆盖。

Case 6 和 Case 11 试图给枚举赋值空字符串或不存在的枚举，这两种情况会导致编译报错，因此没有必要测试。综合以上各种情况，saveMobilePhone 方法一共设计了 9 条手工测试用例。

5.4　TestNG Dubbo 接口自动化测试

在使用 TestNG 进行 Dubbo 接口自动化测试之前，需要先创建一个新的 Maven 项目，关键信息填写如图 5-4 所示。

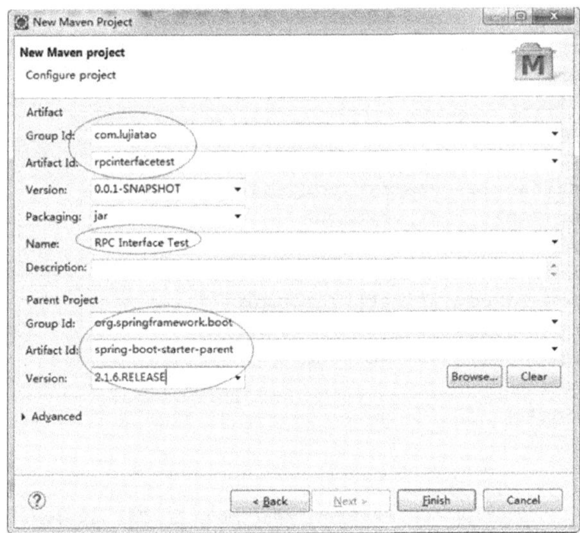

图 5-4

创建完成后，在 pom.xml 文件的<name>标签后输入以下粗体部分内容：

```
<project xmlns="http://maven.apache……"
    xmlns:xsi="http://www.w3……"
    xsi:schemaLocation="http://maven.apache…… http://maven.apache……">
    <modelVersion>4.0.0</modelVersion>

    <parent>
        <groupId>org.springframework.boot</groupId>
        <artifactId>spring-boot-starter-parent</artifactId>
        <version>2.1.6.RELEASE</version>
    </parent>

    <groupId>com.lujiatao</groupId>
    <artifactId>rpcinterfacetest</artifactId>
    <version>0.0.1-SNAPSHOT</version>
    <name>RPC Interface Test</name>

    <dependencies>
        <dependency>
            <groupId>org.testng</groupId>
            <artifactId>testng</artifactId>
            <version>6.14.3</version>
            <scope>test</scope>
        </dependency>
        <dependency>
            <groupId>org.springframework.boot</groupId>
            <artifactId>spring-boot-starter-web</artifactId>
        </dependency>
        <dependency>
```

```
                <groupId>org.apache.dubbo</groupId>
                <artifactId>dubbo</artifactId>
                <version>2.7.1</version>
            </dependency>
            <dependency>
                <groupId>org.apache.curator</groupId>
                <artifactId>curator-recipes</artifactId>
                <version>4.2.0</version>
            </dependency>
            <dependency>
                <groupId>com.alibaba</groupId>
                <artifactId>fastjson</artifactId>
                <version>1.2.59</version>
            </dependency>
        </dependencies>

</project>
```

保存 pom.xml 文件，这时 Maven 会自动下载依赖的 jar 包。这里添加了 5 个依赖。

① TestNG：自动化测试框架 TestNG 的依赖。

② spring-boot-starter-web：作为 Web 开发的基础，这里需要使用其中的一些功能用于测试。

③ Dubbo：分布式微服务框架 Dubbo 的依赖。

④ curator-recipes：注册中心 ZooKeeper 框架的依赖。

⑤ Fastjson：用于将实体类转换为 JSON 字符串，相当于执行了序列化操作。

依赖 jar 包下载完成后，在工程（rpcinterfacetest）上用鼠标右击，从弹出的快捷菜单中选择“TestNG → Convert to TestNG” 选项，在工程中生成 testng.xml 文件。

Dubbo 接口自动化测试的方法有 4 种。

① 基于 XML：使用 XML 文件配置消费者。

② 基于 API：使用 Dubbo API 直接配置消费者。

③ 基于注解：使用注解和 properties 文件配置消费者。

④ 泛化调用：泛化调用主要用于消费者没有 API 接口的场景，这时不需要引入接口的 jar 包，直接通过 GenericService 接口来发起 Dubbo 接口调用，Dubbo 自动化测试框架也是基于泛化调用实现的。

上述方法中的消费者其实就是放置测试用例的工程。

5.4.1 基于 XML 方式

基于 XML、API 和注解的方式均需要依赖接口的 jar 包。从笔者的 GitHub 中下载接口的 jar

包，jar 包名为 mobilephoneservice.jar。下载后在工程（rpcinterfacetest）上用鼠标右击，从弹出的快捷菜单中选择“Build Path → Configure Build Path...” 选项，显示如图 5-5 所示。

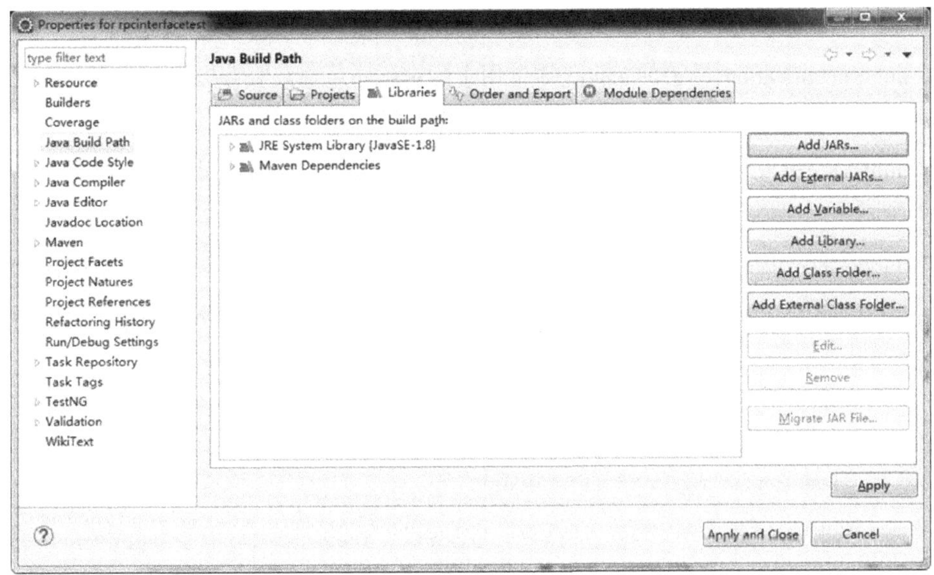

图 5-5

单击“Add External JARs...”选项，选择刚刚下载的“mobilephoneservice.jar”文件，然后单击“Apply and Close”按钮，完成 jar 包的导入。

为简明步骤，采用直接导入 jar 包的方式。在实际项目中，依赖 jar 包一般放在 Maven 私服。将 Eclipse 使用的 Maven 仓库地址配置为 Maven 私服地址。Maven 私服的使用已经超出本书的探讨范围，有兴趣的读者可以自行查阅相关资料。

1. 通过注册中心调用 Dubbo 接口

在 src/test/resources 中创建名为 consumer 的 XML 文件，在 consumer.xml 中输入以下内容：

```
<?xml version="1.0" encoding="UTF-8"?>
<beans xmlns:xsi="http://www.w3……"
    xmlns:dubbo="http://dubbo.apache……"
    xmlns="http://www.springframework……"
    xsi:schemaLocation="http://www.springframework……
        http://www.springframework……
        http://dubbo.apache……
        http://dubbo.apache……">
    <dubbo:application name="RPCDubboConsumer" />
    <dubbo:registry address="zookeeper://localhost:2181" />
    <dubbo:reference id="mobilePhoneService"
```

```
        interface="com.lujiatao.rpcinterface.dubbo.MobilePhoneService"
        version="1.0.0" />
</beans>
```

下面对上述内容进行说明。

① dubbo:application 标签的 name 属性用于指定程序名称，服务提供者和服务消费者的名称不能设置成相同的。

② dubbo:registry 标签的 registry 属性代表配置注册中心，包括注册中心的类型、IP 地址和端口。

③ dubbo:reference 标签的 id 属性用于后续的引用（稍后会介绍），interface 属性设置消费者需要使用的服务，version 属性设置服务版本。需要注意的是，version 如果不填或填错，会造成与服务提供者的接口版本号不一致，从而无法找到服务提供者。

在 src/test/java 中创建名为 com.lujiatao.rpcinterfacetest 的 Package 及名为 GetMobilePhoneTest 的 Class，在 GetMobilePhoneTest 中输入以下内容：

```
package com.lujiatao.rpcinterfacetest;

import org.springframework.context.support.ClassPathXmlApplicationContext;
import org.testng.Assert;
import org.testng.annotations.AfterClass;
import org.testng.annotations.BeforeClass;
import org.testng.annotations.Test;

import com.alibaba.fastjson.JSON;
import com.lujiatao.rpcinterface.domain.MobilePhone;
import com.lujiatao.rpcinterface.dubbo.MobilePhoneService;

public class GetMobilePhoneTest {

    private ClassPathXmlApplicationContext context;
    private MobilePhoneService mobilePhoneService;

    @BeforeClass
    public void init() {
        context = new ClassPathXmlApplicationContext("consumer.xml");
        context.start();
        mobilePhoneService = (MobilePhoneService)
            context.getBean("mobilePhoneService");
    }

    @Test
    public void testCase1() {
        Assert.assertEquals("{\"brand\":\"Apple\",\"model\":\"iPhone
           6S\",\"os\":\"IOS\"}",
                invokeGetMobilePhoneMethod("iPhone 6S"));
    }
```

```
    @Test
    public void testCase2() {
        Assert.assertEquals("null", invokeGetMobilePhoneMethod(""));
    }

    @Test
    public void testCase3() {
        Assert.assertEquals("null", invokeGetMobilePhoneMethod(null));
    }

    @Test
    public void testCase4() {
        Assert.assertEquals("null",
            invokeGetMobilePhoneMethod(
            "01234567890123456789012345678901234567890123456789"));
    }

    private String invokeGetMobilePhoneMethod(String model) {
        MobilePhone mobilePhone = mobilePhoneService.getMobilePhone(model);
        return JSON.toJSONString(mobilePhone);
    }

    @AfterClass
    public void clear() {
        context.close();
    }

}
```

修改 testng.xml 文件，在<test>标签中加入以下粗体部分内容：

```
<?xml version="1.0" encoding="UTF-8"?>
<!DOCTYPE suite SYSTEM "http://testng.org/testng-1.0.dtd">
<suite name="Suite">
    <test thread-count="5" name="Test">
        <classes>
            <class name="com.lujiatao.rpcinterfacetest.GetMobilePhoneTest" />
        </classes>
    </test> <!-- Test -->
</suite> <!-- Suite -->
```

保存所做的修改，在“testng.xml”上用鼠标右击，在弹出的快捷菜单中选择“Run As → TestNG Suite”，然后查看测试报告，如图 5-6 所示。

GetMobilePhoneTest.java　testng.xml　TestNG Report

file:///D:/Data/eclipse/rpcinterfacetest/test-output/emailable-report.html

Test	# Passed	# Skipped	# Failed	Time (ms)	Included Groups	Excluded Groups
Suite						
Test	4	0	0	2,744		

Class	Method	Start	Time (ms)
Suite			
Test — passed			
com.lujiatao.rpcinterfacetest.GetMobilePhoneTest	[illegible]	1565359077167	425
	[illegible]	1565359077595	7
	[illegible]	1565359077605	5
	[illegible]	1565359077612	5

Test

com.lujiatao.rpcinterfacetest.GetMobilePhoneTest#testCase1

图 5-6

下面对上述运行结果进行说明。

① 在 init()方法中，传入 xml 文件作为参数，构建一个 IoC（Inversion of Control，控制反转）容器，并通过 start()方法启动，在 clear()方法中关闭。init()方法中还做了另一件事，就是通过 dubbo:reference 标签的 id 属性值获取服务。

② 在测试用例方法（testCase1—testCase4）中通过封装的 invokeGetMobilePhoneMethod()方法来调用接口中的方法，并对返回值进行断言。这里使用了 Fastjson 中的 toJSONString()方法将实体类转换成了 JSON 字符串。

这里不再演示 saveMobilePhone 的自动化测试用例编写，读者可以按照 getMobilePhone 的示例自行编写，编写方法是一样的。

2. 通过直连方式调用 Dubbo 接口

前面是通过注册中心调用 Dubbo 接口的，其实可以绕过注册中心，直接连接服务提供者来调用 Dubbo 接口。直连的优点是只关心 Dubbo 本身的功能，更为纯粹。缺点是无法模拟真实服务消费者的调用，因为真实服务消费者是通过注册中心的订阅/发布策略来获取服务提供者，并进行 Dubbo 接口调用的。

用直连方式调用 Dubbo 接口很简单，删除 consumer.xml 中的 dubbo:registry 标签，并在 dubbo:reference 标签中配置直连地址即可，如下粗体部分内容所示：

```
<?xml version="1.0" encoding="UTF-8"?>
<beans xmlns:xsi="http://www.w3……"
    xmlns:dubbo="http://dubbo.apache……"
    xmlns="http://www.springframework……"
    xsi:schemaLocation="http://www.springframework……
        http://www.springframework……
```

```
        http://dubbo.apache……
        http://dubbo.apache……">
    <dubbo:application name="RPCDubboConsumer" />
    <dubbo:reference id="mobilePhoneService"
        interface="com.lujiatao.rpcinterface.dubbo.MobilePhoneService"
        version="1.0.0" url="dubbo://localhost:20880" />
</beans>
```

需要说明的是，不管是 ZooKeeper 还是直连端口，都需要根据实际项目而定，这里的端口是待测程序（示例项目）的端口。

5.4.2 基于 API 方式

API 与 XML 有一对一的关系，比如 ApplicationConfig.setName("MyName") 对应 <dubbo:application name="MyName" />。

本节只演示一条自动化测试用例（getMobilePhone 中的 Case 1）的编写，读者在学会方法后，可以自行将所有手工测试用例转化为自动化测试用例。

清空 GetMobilePhoneTest 中的内容，重新输入以下代码：

```
package com.lujiatao.rpcinterfacetest;

import org.apache.dubbo.config.ApplicationConfig;
import org.apache.dubbo.config.ReferenceConfig;
import org.apache.dubbo.config.RegistryConfig;
import org.testng.Assert;
import org.testng.annotations.BeforeClass;
import org.testng.annotations.Test;

import com.alibaba.fastjson.JSON;
import com.lujiatao.rpcinterface.domain.MobilePhone;
import com.lujiatao.rpcinterface.dubbo.MobilePhoneService;

public class GetMobilePhoneTest {

    private MobilePhoneService mobilePhoneService;

    @BeforeClass
    public void init() {
        ApplicationConfig application = new ApplicationConfig();
        application.setName("RPCDubboConsumer");
        RegistryConfig registry = new RegistryConfig();
        registry.setAddress("zookeeper://localhost:2181");
        ReferenceConfig<MobilePhoneService> reference = new
            ReferenceConfig<MobilePhoneService>();
        reference.setApplication(application);
        reference.setRegistry(registry);
        reference.setInterface(MobilePhoneService.class);
```

```
        reference.setVersion("1.0.0");
        mobilePhoneService = reference.get();
    }

    @Test
    public void testCase1() {
        MobilePhone mobilePhone = mobilePhoneService.getMobilePhone("iPhone 6S");
        Assert.assertEquals("{\"brand\":\"Apple\",\"model\":\"iPhone
            6S\",\"os\":\"IOS\"}",
                JSON.toJSONString(mobilePhone));
    }

}
```

下面对上述代码进行说明。

① 在 init()方法中通过 Dubbo API 设置程序名称、注册中心和接口，这里设置的项目和 5.4.1 节中的基本相同，只不过把 XML 换成了 API。

② 删了 clear()方法，因为这里不使用 IoC 读取 XML 文件，因此不需要关闭 ClassPathXmlApplicationContext 对象。

5.4.3 基于注解方式

注解方式比 XML 和 API 两种方式要复杂一些。

在 com.lujiatao.rpcinterfacetest Package 中创建名为 AnnotationClass 和 ConsumerConfiguration 的 Class，输入以下代码：

AnnotationClass：

```
package com.lujiatao.rpcinterfacetest;

import org.apache.dubbo.config.annotation.Reference;
import org.springframework.stereotype.Component;

import com.lujiatao.rpcinterface.domain.MobilePhone;
import com.lujiatao.rpcinterface.dubbo.MobilePhoneService;

@Component("annotationClass")
public class AnnotationClass {

    @Reference(version = "1.0.0")
    private MobilePhoneService mobilePhoneService;

    public MobilePhone getMobilePhone(String model) {
        return mobilePhoneService.getMobilePhone(model);
    }

}
```

该类的作用充当了基于 XML 中 dubbo:reference 标签的角色，即提供服务引用。@Component 注解告诉 Spring 需要为该类创建 Bean，并指定了名称。而@Reference 注解代表引用服务提供者所提供的服务，并指定了版本号。

ConsumerConfiguration：

```
package com.lujiatao.rpcinterfacetest;

import org.apache.dubbo.config.spring.context.annotation.EnableDubbo;
import org.springframework.context.annotation.ComponentScan;
import org.springframework.context.annotation.Configuration;
import org.springframework.context.annotation.PropertySource;

@ComponentScan(value = { "com.lujiatao.rpcinterfacetest" })
@Configuration
@EnableDubbo(scanBasePackages = "com.lujiatao.rpcinterfacetest")
@PropertySource("application.properties")
public class ConsumerConfiguration {
}
```

该类用作载入 Bean 扫描配置、Dubbo 扫描配置和 properties 文件配置。

在 src/test/resources 中创建名为 application.properties 的文件，在 application.properties 中输入以下内容：

```
dubbo.application.name=RPCDubboConsumer
dubbo.registry.address=zookeeper://localhost:2181
```

该文件为配置文件，这里配置了程序名称和注册中心。

清空 GetMobilePhoneTest 中的内容，并重新输入以下代码：

```
package com.lujiatao.rpcinterfacetest;

import
org.springframework.context.annotation.AnnotationConfigApplicationContext;
import org.testng.Assert;
import org.testng.annotations.AfterClass;
import org.testng.annotations.BeforeClass;
import org.testng.annotations.Test;

import com.alibaba.fastjson.JSON;
import com.lujiatao.rpcinterface.domain.MobilePhone;

public class GetMobilePhoneTest {

    private AnnotationConfigApplicationContext context;
    private AnnotationClass annotationClass;

    @BeforeClass
    public void init() {
        context = new
```

```
AnnotationConfigApplicationContext(ConsumerConfiguration.class);
        context.start();
        annotationClass = (AnnotationClass) context.getBean("annotationClass");
    }

    @Test
    public void testCase1() {
        MobilePhone mobilePhone = annotationClass.getMobilePhone("iPhone 6S");
        Assert.assertEquals("{\"brand\":\"Apple\",\"model\":\"iPhone
           6S\",\"os\":\"IOS\"}",
                JSON.toJSONString(mobilePhone));
    }

    @AfterClass
    public void clear() {
        context.close();
    }

}
```

在 init()方法中传入配置类 ConsumerConfiguration，将它作为参数，构建一个 IoC 容器，并通过 start()方法启动，在 clear()方法中关闭。测试用例（testCase1）的写法保持不变。

5.4.4 泛化调用

泛化调用的核心在于使用 GenericService 接口，通过 GenericService 接口的$invoke()方法调用目标 Dubbo 接口。$invoke()方法有 3 个参数，含义如下。

① String method：方法名。

② String[] parameterTypes：参数类型，这里的参数类型需要写全路径，比如“java.lang.String”。

③ Object[] args：参数。

清空 GetMobilePhoneTest 中的内容，并重新输入以下代码：

```
package com.lujiatao.rpcinterfacetest;

import org.apache.dubbo.config.ApplicationConfig;
import org.apache.dubbo.config.ReferenceConfig;
import org.apache.dubbo.config.RegistryConfig;
import org.apache.dubbo.rpc.service.GenericService;
import org.testng.Assert;
import org.testng.annotations.BeforeClass;
import org.testng.annotations.Test;

import com.alibaba.fastjson.JSON;

public class GetMobilePhoneTest {

    private GenericService genericService;
```

```
    @BeforeClass
    public void init() {
        ApplicationConfig application = new ApplicationConfig();
        application.setName("RPCDubboConsumer");
        RegistryConfig registry = new RegistryConfig();
        registry.setAddress("zookeeper://localhost:2181");
        ReferenceConfig<GenericService> reference = new
            ReferenceConfig<GenericService>();
        reference.setApplication(application);
        reference.setRegistry(registry);
        reference.setInterface("com.lujiatao.rpcinterface.dubbo.
            MobilePhoneService");
        reference.setVersion("1.0.0");
        reference.setGeneric(true);
        genericService = reference.get();
    }

    @Test
    public void testCase1() {
        Object mobilePhone = genericService.$invoke("getMobilePhone",
            new String[] { "java.lang.String" },
                new Object[] { "iPhone 6S" });
        Assert.assertEquals("{\"brand\":\"Apple\",\"model\":\"iPhone
            6S\",\"os\":\"IOS\"}",
                JSON.toJSONString(mobilePhone));
    }

}
```

保存所做的修改，在 testng.xml 上用鼠标右击，从弹出的快捷菜单中选择“Run As → TestNG Suite”选项，查看测试报告，如图 5-7 所示。

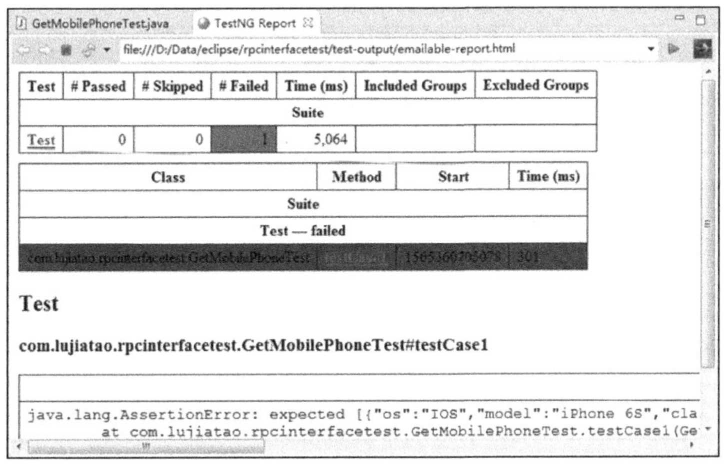

图 5-7

从测试报告中可以看出，测试用例执行失败了。单击“testCase1” 选项，查看失败原因，如图 5-8 所示。

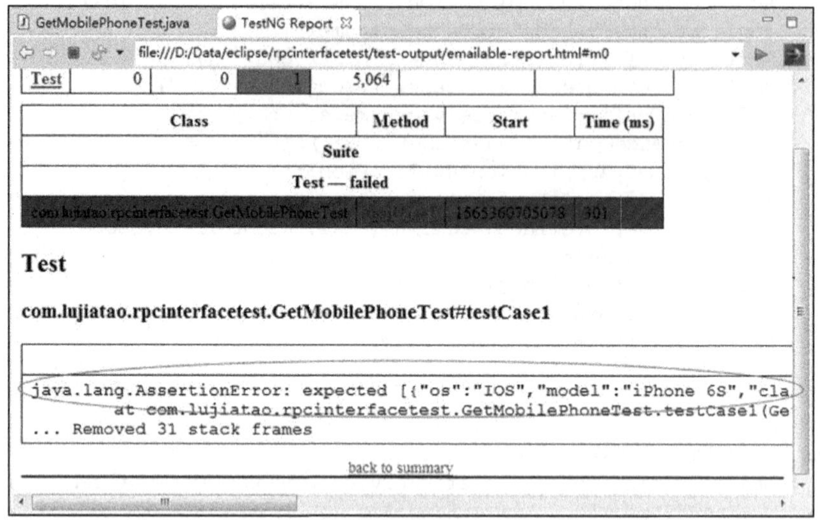

图 5-8

复制 expected 后面的方括号中的内容，如下所示：

```
{
    "os": "IOS",
    "model": "iPhone 6S",
    "class": "com.lujiatao.rpcinterface.domain.MobilePhone",
    "brand": "Apple"
}
```

可以看到方法返回的结果与预期结果有两点不同。

① 返回的字段多了一个 class。为什么会这样呢？因为返回值是一个 MobilePhone 对象，当泛化调用采用 Map 方式表示该对象时，如果不加入 class 这个 Key，则无法判断该对象是哪个类的实例。

② 字段的顺序与预期结果不一致。

如何处理上面两个问题呢？

首先引入一个新的依赖 Gson，即在 pom.xml 文件的<dependencies>标签中加入以下依赖。

```
<dependency>
    <groupId>com.google.code.gson</groupId>
    <artifactId>gson</artifactId>
</dependency>
```

保存 pom.xml 文件，这时 Maven 会自动下载依赖的 jar 包。

然后修改 GetMobilePhoneTest 中的代码，如下面粗体部分内容所示。

```
package com.lujiatao.rpcinterfacetest;

import org.apache.dubbo.config.ApplicationConfig;
import org.apache.dubbo.config.ReferenceConfig;
import org.apache.dubbo.config.RegistryConfig;
import org.apache.dubbo.rpc.service.GenericService;
import org.testng.Assert;
import org.testng.annotations.BeforeClass;
import org.testng.annotations.Test;

import com.alibaba.fastjson.JSON;
import com.google.gson.JsonObject;
import com.google.gson.JsonParser;

public class GetMobilePhoneTest {

    private GenericService genericService;

    @BeforeClass
    public void init() {
        ApplicationConfig application = new ApplicationConfig();
        application.setName("RPCDubboConsumer");
        RegistryConfig registry = new RegistryConfig();
        registry.setAddress("zookeeper://localhost:2181");
        ReferenceConfig<GenericService> reference = new
            ReferenceConfig<GenericService>();
        reference.setApplication(application);
        reference.setRegistry(registry);

    reference.setInterface("com.lujiatao.rpcinterface.dubbo.MobilePhoneService");
        reference.setVersion("1.0.0");
        reference.setGeneric(true);
        genericService = reference.get();
    }

    @Test
    public void testCase1() {
        Object mobilePhone = genericService.$invoke("getMobilePhone",
            new String[] { "java.lang.String" },
                new Object[] { "iPhone 6S" });
        JsonObject expected = (JsonObject) new JsonParser().parse(
                "{\"brand\":\"Apple\",\"model\":\"iPhone
                6S\",\"os\":\"IOS\",\"class\":\"com.lujiatao.rpcinterface.
                domain.MobilePhone\"}");
        JsonObject actual = (JsonObject) new
           JsonParser().parse(JSON.toJSONString(mobilePhone));
        if (!expected.equals(actual)) {
            Assert.fail("失败！");
        }
```

```
        }

    }
```

在预期结果中加入字段 class，引入 gson，将预期结果和实际结果都转换为 JsonObject 对象，并将两个对象进行比较。

保存所做的修改，在“testng.xml”上用鼠标右击，从弹出的快捷菜单中选择“Run As → TestNG Suite”选项，查看测试报告，如图 5-9 所示，表示测试用例执行成功。

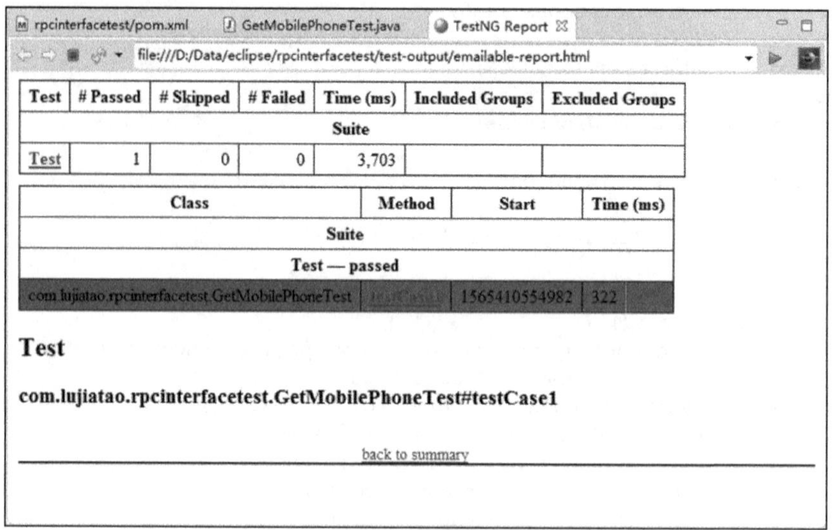

图 5-9

第6章 Web自动化测试

6.1 Web自动化测试工具（框架）简介

Web 作为软件产品中最重要的呈现形式之一，在自动化测试领域中累积了大量的自动化测试工具（框架），常见的如下。

① Selenium：Selenium 是一款免费且开源的 Web 自动化测试工具。Selenium 包括 Selenium IDE、Selenium WebDriver、Selenium Grid 和 Selenium Remote Control 4 个项目，支持 Chrome、Edge、IE、Firefox、Safari 和 Opera 等浏览器的 Web 自动化测试，官方提供对 Java、Python、C#、Ruby 和 JavaScript（Node.js）的支持。由于 Selenium 的开源性，许多公司自研的 Web 自动化测试工具（框架）都集成或扩展至 Selenium。

② UFT：UFT 的英文全称是 Unified Functional Testing，其前身是 QTP（QuickTest Professional），一款由惠普开发的商业自动化测试工具，后来惠普将软件部门卖给了 Micro Focus，因此UFT 目前的持有者是 Micro Focus。UFT 支持 Windows、Android 和 iOS 跨平台，以及 Chrome、IE、Firefox 和 Safari 跨浏览器的自动化测试。

③ TestComplete：TestComplete 是一款由 SmartBear 开发的商业自动化测试工具，支持 Web、Android、iOS 和 Desktop 程序的自动化测试，支持数据驱动和关键字驱动，并提供了详细的测试报告及结果分析功能。

④ Katalon Studio：Katalon Studio 是一款免费但不开源的自动化测试工具，由 Katalon 公司开发，支持 Web、Android 和 iOS 的自动化测试，支持与 Jenkins、Git 和 JIRA 等集成，测试报告包括日志、截图和视频等，非常的多样化。

本章使用 Selenium 作为 Web 自动化测试的工具（框架），并通过集成到 TestNG 来提高自动化测试用例的执行和管理效率。

6.2　部署待测程序

1．安装 JDK

在即将部署待测程序的服务器上安装 JDK。

2．部署待测程序

从笔者的 GitHub 主页下载待测程序，程序名为 webapplication-0.0.1-SNAPSHOT.jar，下载后放在服务器上，执行以下命令运行即可。

```
java -jar E:\webapplication-0.0.1-SNAPSHOT.jar
```

这里笔者把本地电脑作为服务器，且待测程序放在了 E 盘根目录，读者需根据实际情况替换以上路径。运行成功后如图 6-1 所示。

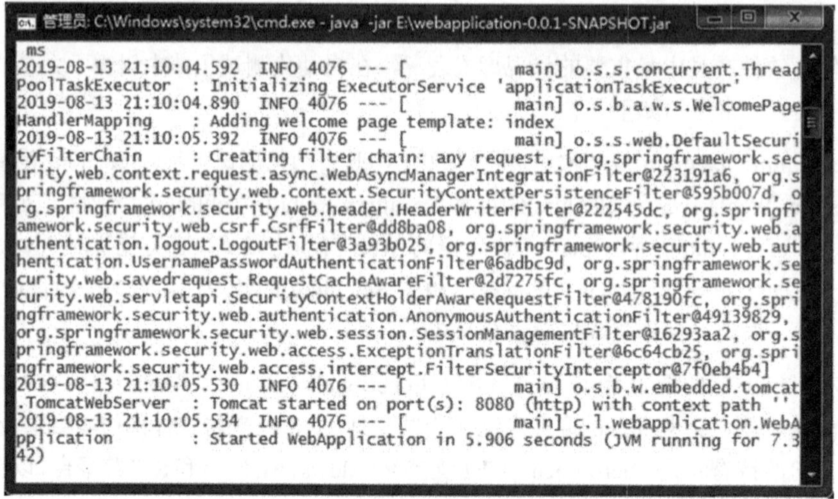

图 6-1

打开浏览器，访问 http://localhost:8080/index-demo，如图 6-2 所示。

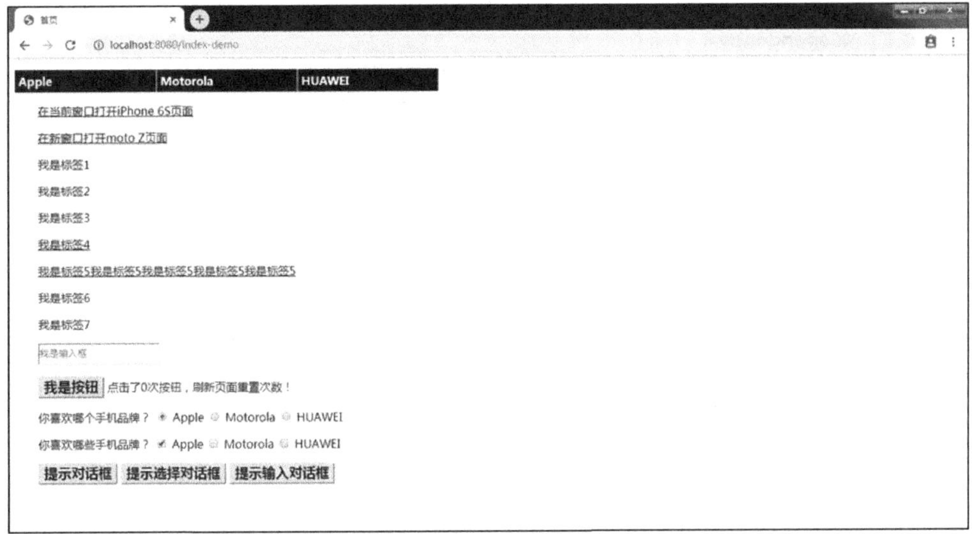

图 6-2

6.3 Selenium 用法

6.3.1 准备

1. 工程创建

在学习 Selenium 之前先创建一个新的 Maven 项目，关键信息填写如下。

Group Id：com.lujiatao

Artifact Id：webapplicationtest

Name：Web Application Test

当然，读者可根据实际情况填写，不需要和笔者填写的完全一致。

创建完成后，在 pom.xml 文件的<name>标签后输入以下粗体部分内容：

```
<project xmlns="http://maven.apache……"
    xmlns:xsi="http://www.w3……"
    xsi:schemaLocation="http://maven.apache…… http://maven.apache……">
    <modelVersion>4.0.0</modelVersion>

    <groupId>com.lujiatao</groupId>
    <artifactId>webapplicationtest</artifactId>
    <version>0.0.1-SNAPSHOT</version>
    <name>Web Application Test</name>
```

```
    <dependencies>
        <dependency>
            <groupId>org.seleniumhq.selenium</groupId>
            <artifactId>selenium-java</artifactId>
            <version>3.141.59</version>
        </dependency>
    </dependencies>

</project>
```

保存 pom.xml 文件，这时 Maven 会自动下载 Selenium 及其依赖的其他 jar 包。

2. 安装浏览器驱动

由于 Selenium 使用了 WebDriver 协议和浏览器通信来实现自动化测试，因此需要一个遵循 WebDriver 协议的服务器充当“中介”，将自动化测试用例通过服务器发送给浏览器执行。在 Web 自动化测试中，这个“中介”我们称之为浏览器驱动，不同的浏览器对应不同的驱动。

如果从 Selenium 官方提供的地址无法下载浏览器驱动，则可以直接在百度中搜索并下载。

以 Chrome 为例，首先下载对应版本的浏览器驱动，下载后为一个压缩文件 chromedriver_win32.zip。然后将压缩文件中的 chromedriver.exe 解压缩到 C:\Windows\System32 目录即可。

6.3.2 元素操作

Web 页面中有输入框、按钮、单选框和复选框等，在 Web 自动化测试中，这些统称为 Web 元素（以下简称元素）。在对元素进行操作之前，需要先定位元素。

1. 定位元素

Selenium 提供了 8 种定位元素的方法。

（1）ID

在同一个页面中，元素 ID 具有唯一性，因此通过元素 ID 定位出来的元素具有唯一性。在 CSS 中用井号（#）表示对 ID 的引用。

（2）Name

在同一个页面中，元素 Name 不具有唯一性，因此通过元素 Name 定位出来的元素可能有多个。在表单中经常使用 Name 作为数据传输的参数名。

（3）Class

Class 为元素的类，在 CSS 中用英文句号（.）表示对类的引用。顾名思义，“类”表示多个元素具备一类特性（通常是 CSS 的特性）。

（4）Tag Name

在 HTML 中有很多标签，比如<p>、<a>和<span>等，可以通过标签名来定位元素。

（5）Link Text

在 Web 页面中经常会出现超链接，可以通过超链接的文本对元素进行定位。

（6）Partial Link Text

和 Link Text 类似，只不过有些超链接的文本太长，写起来比较麻烦，这时可以考虑使用 Partial Link Text，即只通过超链接的部分文本对元素进行定位。

（7）XPath

XPath（XML Path Language，XML 路径语言）通过 HTML 的路径来定位元素。路径可以是绝对路径，也可以是相对路径。

（8）CSS Selector

CSS Selector （CSS 选择器）可以通过类、ID 和标签等定位元素。

由于以上定位方法可能获取到一个或多个元素，因此 Selenium 提供了两种方法获取元素，即 findElement()和 findElements()，这两种方法都是通过传递一个 By 对象来定位元素的。在 By 对象中，再调用具体方法来定位元素。

2．操作元素

操作元素的方式很多，常见的比如点单击、输入文本、获取文本和获取属性等。

下面举例说明定位元素和操作元素。

在 src/test/java 中创建名为 com.lujiatao.webapplicationtest 的 Package 及名为 SeleniumTest 的 Class（需要勾选“public static void main(String[] args)”），在 SeleniumTest 中输入以下代码：

```
package com.lujiatao.webapplicationtest;

import org.openqa.selenium.By;
import org.openqa.selenium.WebDriver;
import org.openqa.selenium.WebElement;
import org.openqa.selenium.chrome.ChromeDriver;

public class SeleniumTest {

    public static void main(String[] args) {
        WebDriver chromeDriver = new ChromeDriver();
        chromeDriver.get("http://localhost:8080/index-demo");
        // 定位元素
        WebElement e1 = chromeDriver.findElement(By.id("label-id"));
        System.out.println("通过 ID 定位获取的元素为：" + e1.getText());
```

```
        WebElement e2 = chromeDriver.findElement(By.name("label-name"));
        System.out.println("通过 Name 定位获取的元素为：" +
            e2.getAttribute("placeholder"));

        WebElement e3 = chromeDriver.findElement(By.className("label-class"));
        System.out.println("通过 Class 定位获取的元素为：" + e3.getText());

        WebElement e4 = chromeDriver.findElements(By.tagName("label")).get(2);
        System.out.println("通过 Tag Name 定位获取的元素为：" + e4.getText());

        WebElement e5 = chromeDriver.findElement(By.linkText("我是标签 4"));
        System.out.println("通过 Link Text 定位获取的元素为：" + e5.getText());

        WebElement e6 = chromeDriver.findElement(By.partialLinkText("我是标签 5"));
        System.out.println("通过 Partial Link Text 定位获取的元素为：" + e6.getText());

        WebElement e7 = chromeDriver.findElement(By.xpath("//div/p[8]/label"));
        System.out.println("通过 XPath 定位获取的元素为：" + e7.getText());

        WebElement e8 = chromeDriver.findElement(By.cssSelector("div >
           p:nth-child(9) > label"));
        System.out.println("通过 CSS Selector 定位获取的元素为：" + e8.getText());
        // 操作元素
        e2.sendKeys("我输入了内容");
        WebElement e9 = chromeDriver.findElement(By.id("my-button"));
        e9.click();
        e9.click();
    }

}
```

保存代码，按快捷键 F11 运行工程，此时 Eclipse 的控制台输出如下：

```
Starting ChromeDriver 76.0.3809.68
(420c9498db8ce8fcd190a954d51297672c1515d5-refs/branch-heads/3809@{#864}) on port
29666
Only local connections are allowed.
Please protect ports used by ChromeDriver and related test frameworks to prevent
access by malicious code.
八月 14, 2019 10:17:08 下午 org.openqa.selenium.remote.ProtocolHandshake
createSession
信息: Detected dialect: W3C
通过 ID 定位获取的元素为：我是标签 1
通过 Name 定位获取的元素为：我是输入框
通过 Class 定位获取的元素为：我是标签 2
通过 Tag Name 定位获取的元素为：我是标签 3
通过 Link Text 定位获取的元素为：我是标签 4
```

```
通过 Partial Link Text 定位获取的元素为：我是标签 5 我是标签 5 我是标签 5 我是标签 5 我是标签 5
通过 XPath 定位获取的元素为：我是标签 6
通过 CSS Selector 定位获取的元素为：我是标签 7
```

Chrome 浏览器会自动打开 index-demo 页面，并显示如图 6-3 所示的内容。

图 6-3

下面对上述运行结果进行说明。

① 在对元素进行操作前需要先打开浏览器，这里使用了 new ChromeDriver()创建一个浏览器驱动对象，然后调用其中的 get()方法，并传入 URL 作为参数导航至 index-demo 页面。

② 当使用 Tag Name 定位元素时，由于该页面有多个 label 元素，所以这里使用了 findElements()方法，该方法可返回多个元素，用 get()方法获取其中一个元素（这里获取的是下标为 2 的元素，即第 3 个元素）。如何知道页面有多个 label 元素呢？在 Chrome 中按快捷键 F12，单击“Elements”标签查看页面源码，可以看到页面有多个 label 元素，如图 6-4 所示。

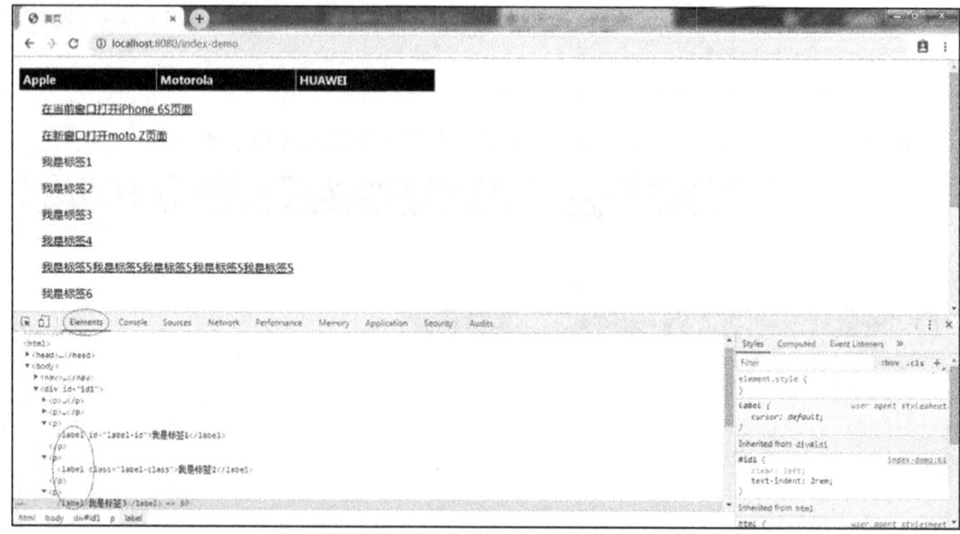

图 6-4

通过这种方法还可以查看元素的其他信息，比如 ID、name 和 class 等。

③ 在示例中，XPath 表达式的写法为“//div/p[8]/label”。其中，“//”代表相对路径，如果要用绝对路径，则需要从页面的根元素开始书写，即“html/body/div/p[8]/label”。p[8]代表第 8 个<p>标签。

④ 在示例中，CSS 选择器的写法为“div > p:nth-child(9) > label”。其中，“>”代表前者是后者的父元素，而“:nth-child(9)”代表第 9 个子元素，这里指第 9 个<p>标签。

⑤ 通过自动化测试代码默认打开的浏览器并不是全屏的，如果需要全屏显示，可以通过调用浏览器驱动对象中的 manage().window().maximize()方法将浏览器最大化。

⑥ 在获取到元素后，通过 getText()方法获取元素的文本，通过 getAttribute()方法获取元素的属性值，该方法的参数为属性名。

⑦ 从图 6-3 可以看出，在输入框中显示的文本为“我输入了内容”，这是通过调用 sendKeys()方法传入的文本。

⑧ 从图 6-3 可以看出，在“我是按钮”旁显示了“点击了 2 次按钮，刷新页面重置次数！”，这是通过两次调用 click()方法点击了两次该按钮。

⑨ 在 Eclipse 控制台中开头有一段文字，这段文字说明了 Chrome 浏览器驱动启动后监听的端口以及安全等方面的信息。

3．遍历元素

在实际项目中经常需要获取一组元素，然后对元素进行遍历，找到满足条件的元素。

下面就遍历一组单选框，并找到其中已选中的那个。

删除 SeleniumTest 中的内容，输入以下代码：

```
package com.lujiatao.webapplicationtest;

import java.util.List;

import org.openqa.selenium.By;
import org.openqa.selenium.WebDriver;
import org.openqa.selenium.WebElement;
import org.openqa.selenium.chrome.ChromeDriver;

public class SeleniumTest {

    public static void main(String[] args) {
        WebDriver chromeDriver = new ChromeDriver();
        chromeDriver.get("http://localhost:8080/index-demo");
        List<WebElement> radios = chromeDriver.findElements(By.name("radio"));
        for (int i = 0; i < radios.size(); i++) {
            if (radios.get(i).isSelected()) {
                System.out.println("第" + (i + 1) + "个单选框被选中了!");
            }
        }
    }

}
```

保存代码，按快捷键 F11 运行工程，此时 Eclipse 的控制台输出如下。

```
Starting ChromeDriver 76.0.3809.68
(420c9498db8ce8fcd190a954d51297672c1515d5-refs/branch-heads/3809@{#864}) on port
1320
Only local connections are allowed.
Please protect ports used by ChromeDriver and related test frameworks to prevent
access by malicious code.
八月 16, 2019 9:52:19 下午 org.openqa.selenium.remote.ProtocolHandshake
createSession
信息: Detected dialect: W3C
第 1 个单选框被选中了!
```

从 index-demo 页面可以看出，第 1 个单选框处于选中状态，控制台打印的结果与预期一致。

6.3.3　鼠标事件

在鼠标事件中最常见的就是单击，单击是通过调用浏览器驱动对象的 click()方法实现的。另外，鼠标还有很多其他事件，这些事件需要通过 Actions 对象进行调用，即在创建 Actions 对象时，将浏览器驱动对象作为参数传入。常见的鼠标事件如表 6-1 所示。

表 6-1

事件名称	对应方法	说　　明
单击	click()	单击鼠标左键，其效果和浏览器驱动对象中的click()方法的效果一致
双击	doubleClick()	双击鼠标左键
右击	contextClick()	单击鼠标右键
按下	clickAndHold()	按下鼠标左键，但不松开
松开	release()	释放鼠标
拖动	dragAndDrop()	拖动元素A到元素B（或到某个坐标点）
悬停	moveToElement()	悬停在某个元素上方

通过 Actions 对象调用这些方法后并没有真正执行，可理解为将要执行的操作先存储起来，最后再调用 perform()方法执行。

下面以鼠标悬停 moveToElement()方法为例，介绍鼠标事件。

打开 Chrome 浏览器，访问 http://localhost:8080/index-demo 页面，将鼠标悬停在页面左上角的“Apple”上，如图 6-5 所示。

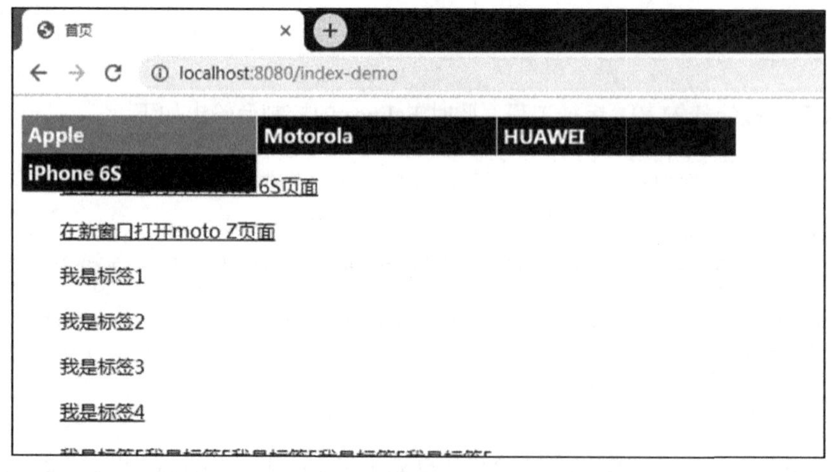

图 6-5

当鼠标悬停在“Apple”上时打开了一个隐藏菜单，在隐藏菜单上用鼠标右击，从弹出的快捷菜单中选择“检查”选项，进入查看页面源码界面（按快捷键 F12 也可以进入该界面）。可以在这里分析元素，以便后续对元素进行操作。

下面通过自动化测试定位到隐藏菜单的“iPhone 6S”元素并单击。

删除 SeleniumTest 中的内容，输入以下代码：

```
package com.lujiatao.webapplicationtest;

import org.openqa.selenium.By;
import org.openqa.selenium.WebDriver;
import org.openqa.selenium.WebElement;
import org.openqa.selenium.chrome.ChromeDriver;
import org.openqa.selenium.interactions.Actions;

public class SeleniumTest {

    public static void main(String[] args) {
        WebDriver chromeDriver = new ChromeDriver();
        chromeDriver.get("http://localhost:8080/index-demo");
        WebElement apple = chromeDriver.findElement(By.xpath("//nav/ul/li[1]/a"));
        WebElement iPhone6S =
            chromeDriver.findElement(By.xpath("//ul/li[1]/ul/li/a"));
        Actions action = new Actions(chromeDriver);
        action.moveToElement(apple);
        action.moveToElement(iPhone6S);
        action.click();
        action.perform();
    }

}
```

保存代码，按快捷键 F11 运行工程，此时 Chrome 浏览器会自动打开 index-demo 页面，如图 6-6 所示。

图 6-6

下面对运行结果进行说明。

① 通过查看页面源码获得“Apple”和“iPhone 6S”这两个元素的 XPath 路径，使用 XPath 定位这两个元素。

② 创建一个 Actions 对象，使用 moveToElement()方法将鼠标先悬停在“Apple”元素上，再悬停在“iPhone 6S”元素上，单击“iPhone 6S”元素。

③ 调用 perform()方法执行上述所有动作，页面会跳转到新的页面，即图 6-6 所示的页面。

6.3.4　键盘事件

键盘事件是通过 sendKeys()方法调用的。在 6.3.2 节的示例中，通过给 sendKeys()方法传入“我输入了内容”作为参数向输入框中输入了文本。其实 sendKeys()方法可以接收很多参数来模拟键盘事件，常见的键盘事件如表 6-2 所示。

表 6-2

事件名称	对应参数	说　明
退格	sendKeys(Keys.BACK_SPACE)	模拟键盘按Backspace键
回车	sendKeys(Keys.ENTER)	模拟键盘按Enter键
全选	sendKeys(Keys.CONTROL, "a")	模拟键盘按Ctrl + a键
复制	sendKeys(Keys.CONTROL, "c")	模拟键盘按Ctrl + c键
剪切	sendKeys(Keys.CONTROL, "x")	模拟键盘按Ctrl + x键
粘贴	sendKeys(Keys.CONTROL, "v")	模拟键盘按Ctrl + v键

下面以退格 sendKeys(Keys.BACK_SPACE)为例介绍键盘事件。

删除 SeleniumTest 中的内容，输入以下代码：

```
package com.lujiatao.webapplicationtest;

import org.openqa.selenium.By;
import org.openqa.selenium.Keys;
import org.openqa.selenium.WebDriver;
import org.openqa.selenium.WebElement;
import org.openqa.selenium.chrome.ChromeDriver;

public class SeleniumTest {

    public static void main(String[] args) {
        WebDriver chromeDriver = new ChromeDriver();
        chromeDriver.get("http://localhost:8080/index-demo");
        WebElement input = chromeDriver.findElement(By.name("label-name"));
        input.sendKeys("123456");
        System.out.println(input.getAttribute("value"));
        input.sendKeys(Keys.BACK_SPACE);
        System.out.println(input.getAttribute("value"));
    }

}
```

保存代码，按快捷键 F11 运行工程，此时 Eclipse 的控制台输出如下：

```
Starting ChromeDriver 76.0.3809.68
(420c9498db8ce8fcd190a954d51297672c1515d5-refs/branch-heads/3809@{#864}) on port
27160
Only local connections are allowed.
Please protect ports used by ChromeDriver and related test frameworks to prevent
```

```
access by malicious code.
八月 16, 2019 8:41:51 下午 org.openqa.selenium.remote.ProtocolHandshake
createSession
信息: Detected dialect: W3C
123456
12345
```

Chrome 浏览器会自动打开 index-demo 页面，并显示如图 6-7 所示的内容。

图 6-7

当定位到输入框后，先在输入框中输入“123456”，然后模拟键盘按退格键删除“6”，因此第二次打印到控制台的字符串为“12345”，页面上也能看到最后显示的是“12345”。

6.3.5 浏览器操作

前面介绍的都是页面级的操作，没有涉及浏览器级别的操作。浏览器级别的常用操作包括前进、后退、刷新、切换窗口、关闭窗口和关闭浏览器等，下面举例说明。

删除 SeleniumTest 中的内容，输入以下代码：

```
package com.lujiatao.webapplicationtest;

import org.openqa.selenium.By;
import org.openqa.selenium.WebDriver;
import org.openqa.selenium.WebDriver.Navigation;
import org.openqa.selenium.WebElement;
import org.openqa.selenium.chrome.ChromeDriver;
```

```
public class SeleniumTest {

    public static void main(String[] args) {
        WebDriver chromeDriver = new ChromeDriver();
        Navigation navigation = chromeDriver.navigate();
        navigation.to("http://localhost:8080/index-demo");
        waitOneSecond();
        navigation.to("http://localhost:8080/openIPhone6S");
        waitOneSecond();
        navigation.back();
        waitOneSecond();
        navigation.forward();
        waitOneSecond();
        navigation.refresh();
        waitOneSecond();
        navigation.back();
        waitOneSecond();
        WebElement motoZ = chromeDriver.findElement(By.linkText("在新窗口打开moto
            Z 页面"));
        motoZ.click();
        waitOneSecond();
        String[] windows = new String[chromeDriver.getWindowHandles().size()];
        chromeDriver.getWindowHandles().toArray(windows);
        chromeDriver.switchTo().window(windows[0]);
        waitOneSecond();
        chromeDriver.close();
        waitOneSecond();
        chromeDriver.switchTo().window(windows[1]);
        waitOneSecond();
        chromeDriver.quit();
    }

    public static void waitOneSecond() {
        try {
            Thread.sleep(1000);
        } catch (InterruptedException e) {
            e.printStackTrace();
        }
    }

}
```

保存代码，按快捷键 F11 运行工程，此时 Chrome 浏览器会自动打开 index-demo 页面，完成一系列操作后自动关闭浏览器。

下面对运行结果进行说明。

① 封装一个静态方法 waitOneSecond()，以便于在每个步骤之间加入等待时间，观察测试代码执行情况，否则运行太快很难看清楚。

② 通过调用浏览器驱动对象的 navigate()方法获取一个导航对象，再通过导航对象进行前进、后退和刷新等操作。

③ 通过调用浏览器驱动对象的 getWindowHandles()方法获取浏览器窗口的句柄，并将其存放入一个数组中，随后进行切换窗口、关闭窗口和关闭浏览器等操作。close()方法可关闭当前窗口，而 quit()方法可关闭整个浏览器。

6.3.6　JavaScript 对话框处理

虽然在 Web 前端开发中通过各种框架封装的模态框十分流行，但 JavaScript 经典的对话框也时有出现，Selenium 处理 JavaScript 对话框的方式是通过调用浏览器驱动对象中的 switchTo().alert()方法返回一个 Alert 对象，再对对话框进行处理的。本节介绍 3 种 JavaScript 对话框及其处理方法。JavaScript 对话框及其处理方法如表 6-3 所示。

表 6-3

JavaScript对话框	Alert对象方法	方法说明
Alert	accept()	单击对话框确定
Alert	getText()	获取对话框文本
Confirm	accept()	单击对话框确定
Confirm	dismiss()	单击对话框取消
Confirm	getText()	获取对话框文本
Prompt	accept()	单击对话框确定
Prompt	dismiss()	单击对话框取消
Prompt	sendKeys()	对话框输入内容
Prompt	getText()	获取对话框文本

这些对话框的外观在不同浏览器中有所区别，但基本要素都是一样的，以 Chrome 浏览器为例，三种对话框的外观如图 6-8 到图 6-10 所示。

图 6-8

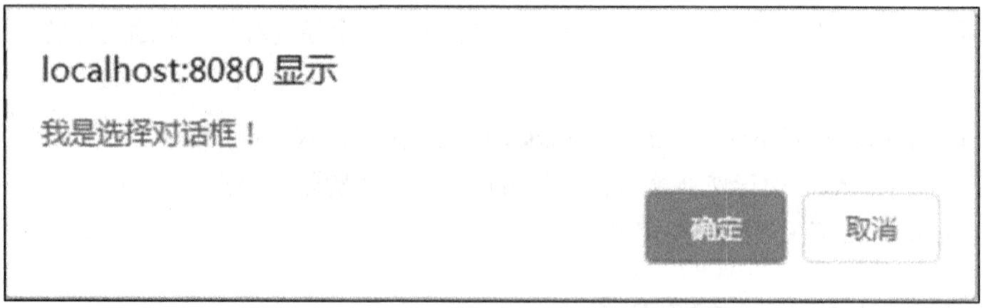

图 6-9

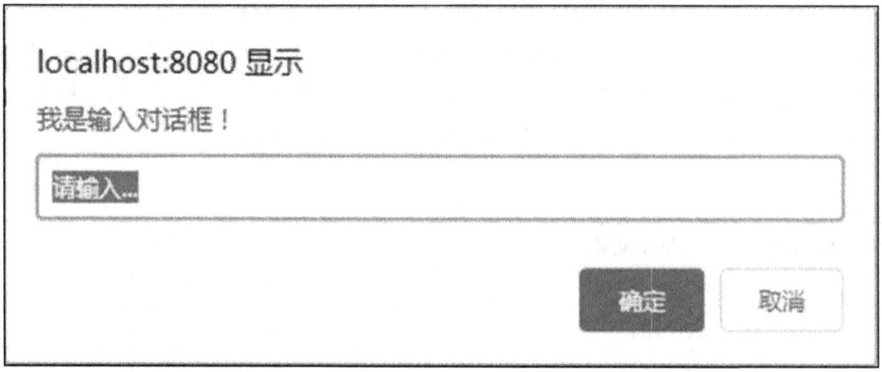

图 6-10

下面通过 Java 代码处理这些对话框。

删除 SeleniumTest 中的内容，输入以下代码：

```
package com.lujiatao.webapplicationtest;

import org.openqa.selenium.Alert;
import org.openqa.selenium.By;
import org.openqa.selenium.WebDriver;
import org.openqa.selenium.chrome.ChromeDriver;

public class SeleniumTest {

    public static void main(String[] args) {
        WebDriver chromeDriver = new ChromeDriver();
        chromeDriver.get("http://localhost:8080/index-demo");
        waitOneSecond();
        // Alert 对话框处理
        chromeDriver.findElement(By.id("my-button2")).click();
        Alert alert = chromeDriver.switchTo().alert();
        System.out.println("Alert 对话框的文本为：" + alert.getText());
        alert.accept();
        waitOneSecond();
        // Confirm 对话框处理
```

```
        chromeDriver.findElement(By.id("my-button3")).click();
        alert = chromeDriver.switchTo().alert();
        System.out.println("Confirm对话框的文本为：" + alert.getText());
        alert.accept();
        waitOneSecond();
        // Prompt对话框处理
        chromeDriver.findElement(By.id("my-button4")).click();
        alert = chromeDriver.switchTo().alert();
        System.out.println("Prompt对话框的文本为：" + alert.getText());
        alert.accept();
        waitOneSecond();
        chromeDriver.quit();
    }

    public static void waitOneSecond() {
        try {
            Thread.sleep(1000);
        } catch (InterruptedException e) {
            e.printStackTrace();
        }
    }

}
```

保存代码，按快捷键 F11 运行工程，此时 Chrome 浏览器会自动打开 index-demo 页面，完成一系列操作后自动关闭浏览器。另外，Eclipse 的控制台输出如下：

```
Starting ChromeDriver 76.0.3809.68
(420c9498db8ce8fcd190a954d51297672c1515d5-refs/branch-heads/3809@{#864}) on port
26375
Only local connections are allowed.
Please protect ports used by ChromeDriver and related test frameworks to prevent
access by malicious code.
八月 17, 2019 11:32:47 上午 org.openqa.selenium.remote.ProtocolHandshake
createSession
信息: Detected dialect: W3C
Alert对话框的文本为：我是提示对话框！
Confirm对话框的文本为：我是选择对话框！
Prompt对话框的文本为：我是输入对话框！
```

下面对运行结果进行说明。

① 首先通过 ID 定位元素，然后单击对应按钮，打开对应的提示框。

② 在获取到 Alert 对象后，获取对话框的文本，之后关闭对话框。

③ 在每个对话框打开和关闭前后都有 1s 的等待时间，便于我们观察测试代码的运行。

④ 最后调用浏览器驱动的 quit()方法关闭浏览器。

6.3.7　等待处理

6.3.6 节中使用 Thread.sleep(1000)强制等待了 1s，在 Selenium 中有更为智能的等待方式，分别为显式等待和隐式等待。

1. 显式等待

显式等待是指在需要等待的位置指定等待时间，Thread.sleep()方法可以看成是显式等待的一种，但它并不智能，因为没有条件来中止等待。这里介绍的 Selenium 的显式等待，它会指定一个超时时间和中止条件，当条件满足时会中止等待。

删除 SeleniumTest 中的内容，输入以下代码：

```
package com.lujiatao.webapplicationtest;

import org.openqa.selenium.By;
import org.openqa.selenium.WebDriver;
import org.openqa.selenium.chrome.ChromeDriver;
import org.openqa.selenium.support.ui.ExpectedConditions;
import org.openqa.selenium.support.ui.WebDriverWait;

public class SeleniumTest {

    public static void main(String[] args) {
        WebDriver chromeDriver = new ChromeDriver();
        chromeDriver.get("http://localhost:8080/index-demo");
        long start = System.currentTimeMillis() / 1000;
        waitMostTenSeconds(chromeDriver, By.id("my-button3"));
        long end = System.currentTimeMillis() / 1000;
        System.out.println("等待了" + (end - start) + "秒！");
        start = System.currentTimeMillis() / 1000;
        waitMostTenSeconds(chromeDriver, By.id("my-button5"));
        end = System.currentTimeMillis() / 1000;
        System.out.println("等待了" + (end - start) + "秒！");
        chromeDriver.quit();
    }

    public static void waitMostTenSeconds(WebDriver chromeDriver, By by) {
        WebDriverWait webDriverWait = new WebDriverWait(chromeDriver, 10);
        try {

    webDriverWait.until(ExpectedConditions.presenceOfElementLocated(by));
        } catch (Exception e) {
            e.printStackTrace();
        }
    }

}
```

保存代码，按快捷键 F11 运行工程，此时 Eclipse 的控制台输出如下：

```
Starting ChromeDriver 76.0.3809.68
(420c9498db8ce8fcd190a954d51297672c1515d5-refs/branch-heads/3809@{#864}) on port
44237
Only local connections are allowed.
Please protect ports used by ChromeDriver and related test frameworks to prevent
access by malicious code.
八月 17, 2019 12:54:40 下午 org.openqa.selenium.remote.ProtocolHandshake
createSession
信息: Detected dialect: W3C
等待了 1 秒!
等待了 10 秒!
org.openqa.selenium.TimeoutException: Expected condition failed: waiting for
presence of element located by: By.id: my-button5 (tried for 10 second(s) with 500
milliseconds interval)
    at
org.openqa.selenium.support.ui.WebDriverWait.timeoutException(WebDriverWait.jav
a:95)
    at org.openqa.selenium.support.ui.FluentWait.until(FluentWait.java:272)
    at
com.lujiatao.webapplicationtest.SeleniumTest.waitMostTenSeconds(SeleniumTest.ja
va:28)
    at com.lujiatao.webapplicationtest.SeleniumTest.main(SeleniumTest.java:19)
Caused by: org.openqa.selenium.NoSuchElementException: no such element: Unable to
locate element: {"method":"css selector","selector":"#my\-button5"}
  (Session info: chrome=76.0.3809.100)
For documentation on this error, please visit:
https://www.seleniumhq.org/exceptions/no_such_element.html
Build info: version: '3.141.59', revision: 'e82be7d358', time:
'2018-11-14T08:17:03'
System info: host: 'LUJIATAO-PC', ip: '192.168.3.12', os.name: 'Windows 7', os.arch:
'amd64', os.version: '6.1', java.version: '1.8.0_212'
Driver info: org.openqa.selenium.chrome.ChromeDriver
Capabilities {acceptInsecureCerts: false, browserName: chrome, browserVersion:
76.0.3809.100, chrome: {chromedriverVersion: 76.0.3809.68 (420c9498db8ce...,
userDataDir: C:\Users\lujiatao\AppData\L...}, goog:chromeOptions:
{debuggerAddress: localhost:7724}, javascriptEnabled: true,
networkConnectionEnabled: false, pageLoadStrategy: normal, platform: XP,
platformName: XP, proxy: Proxy(), setWindowRect: true, strictFileInteractability:
false, timeouts: {implicit: 0, pageLoad: 300000, script: 30000},
unhandledPromptBehavior: dismiss and notify}
Session ID: 8dd43c0ebae0da866dac78a6aab6c14b
*** Element info: {Using=id, value=my-button5}
    at sun.reflect.GeneratedConstructorAccessor12.newInstance(Unknown Source)
    at sun.reflect.DelegatingConstructorAccessorImpl.
newInstance(DelegatingConstructorAccessorImpl.java:45)
    at java.lang.reflect.Constructor.newInstance(Constructor.java:423)
    at org.openqa.selenium.remote.http.W3CHttpResponseCodec.
createException(W3CHttpResponseCodec.java:187)
    at org.openqa.selenium.remote.http.W3CHttpResponseCodec.
decode(W3CHttpResponseCodec.java:122)
```

```
    at org.openqa.selenium.remote.http.W3CHttpResponseCodec.
decode(W3CHttpResponseCodec.java:49)
    at org.openqa.selenium.remote.HttpCommandExecutor.
execute(HttpCommandExecutor.java:158)
    at org.openqa.selenium.remote.service.DriverCommandExecutor.
execute(DriverCommandExecutor.java:83)
    at org.openqa.selenium.remote.RemoteWebDriver.
execute(RemoteWebDriver.java:552)
    at org.openqa.selenium.remote.RemoteWebDriver.
findElement(RemoteWebDriver.java:323)
    at org.openqa.selenium.remote.RemoteWebDriver.
findElementById(RemoteWebDriver.java:372)
    at org.openqa.selenium.By$ById.findElement(By.java:188)
    at org.openqa.selenium.remote.RemoteWebDriver.
findElement(RemoteWebDriver.java:315)
    at org.openqa.selenium.support.ui.ExpectedConditions$6.
apply(ExpectedConditions.java:182)
    at org.openqa.selenium.support.ui.ExpectedConditions$6.
apply(ExpectedConditions.java:179)
    at org.openqa.selenium.support.ui.FluentWait.until(FluentWait.java:249)
    ... 2 more
```

下面对运行结果进行说明。

① 定义 start 和 end 两个变量，存储等待开始和结束的时间。

② 封装一个 waitMostTenSeconds()方法用于等待。该方法通过传入浏览器驱动对象及 By 对象作为参数，方法内使用了 WebDriverWait 对象，并调用了 WebDriverWait 对象的 until()方法，且在 until()方法内传入了期望条件（这里的期望条件是元素被加载）。

③ 因为 ID 为“my-button3”的元素存在，且在打开 index-demo 页面时就被加载了，所以等待时间为 0s。ID 为“my-button5”的元素不存在，因此在等待 10s 后抛出了 TimeoutException（超时）异常。

④ 最后调用浏览器驱动的 quit()方法关闭浏览器。

2. 隐式等待

隐式等待和显式等待的区别在于，隐式等待是全局性的，不需要在某个位置单独指定等待时间，而是有一个全局的设置。隐式等待需要浏览器驱动对象调用 manage().timeouts()方法获得一个 Timeouts 对象，然后再对隐式等待进行定义。在 Timeouts 中有 3 种等待方式，如表 6-4 所示。

表 6-4

隐式等待	Timeouts对象方法	方法说明
查找元素等待	implicitlyWait()	全局设置查找元素超时的时间
载入页面等待	pageLoadTimeout()	全局设置载入页面超时的时间
JavaScript脚本等待	setScriptTimeout()	全局设置JavaScript脚本异步执行超时的时间

这里以查找元素等待 implicitlyWait()为例。

删除 SeleniumTest 中的内容，输入以下代码：

```
package com.lujiatao.webapplicationtest;

import java.util.concurrent.TimeUnit;

import org.openqa.selenium.By;
import org.openqa.selenium.WebDriver;
import org.openqa.selenium.chrome.ChromeDriver;

public class SeleniumTest {

    public static void main(String[] args) {
        WebDriver chromeDriver = new ChromeDriver();
        chromeDriver.manage().timeouts().implicitlyWait(10, TimeUnit.SECONDS);
        chromeDriver.get("http://localhost:8080/index-demo");
        long start = System.currentTimeMillis() / 1000;
        chromeDriver.findElement(By.id("my-button3"));
        long end = System.currentTimeMillis() / 1000;
        System.out.println("等待了" + (end - start) + "秒！");
        start = System.currentTimeMillis() / 1000;
        try {
            chromeDriver.findElement(By.id("my-button5"));
        } catch (Exception e) {
            e.printStackTrace();
        }
        end = System.currentTimeMillis() / 1000;
        System.out.println("等待了" + (end - start) + "秒！");
        chromeDriver.quit();
    }

}
```

保存代码，按快捷键 F11 运行工程，此时 Eclipse 的控制台输出如下：

```
Starting ChromeDriver 76.0.3809.68
(420c9498db8ce8fcd190a954d51297672c1515d5-refs/branch-heads/3809@{#864}) on port
33001
Only local connections are allowed.
Please protect ports used by ChromeDriver and related test frameworks to prevent
access by malicious code.
```

```
八月 17, 2019 3:03:46 下午 org.openqa.selenium.remote.ProtocolHandshake
createSession
信息: Detected dialect: W3C
等待了 0 秒!
org.openqa.selenium.NoSuchElementException: no such element: Unable to locate
element: {"method":"css selector","selector":"#my\-button5"}
  (Session info: chrome=76.0.3809.100)
For documentation on this error, please visit:
https://www.seleniumhq.org/exceptions/no_such_element.html
Build info: version: '3.141.59', revision: 'e82be7d358', time:
'2018-11-14T08:17:03'
System info: host: 'LUJIATAO-PC', ip: '192.168.3.12', os.name: 'Windows 7', os.arch:
'amd64', os.version: '6.1', java.version: '1.8.0_212'
Driver info: org.openqa.selenium.chrome.ChromeDriver
Capabilities {acceptInsecureCerts: false, browserName: chrome, browserVersion:
76.0.3809.100, chrome: {chromedriverVersion: 76.0.3809.68 (420c9498db8ce...,
userDataDir: C:\Users\lujiatao\AppData\L...}, goog:chromeOptions:
{debuggerAddress: localhost:13081}, javascriptEnabled: true,
networkConnectionEnabled: false, pageLoadStrategy: normal, platform: XP,
platformName: XP, proxy: Proxy(), setWindowRect: true, strictFileInteractability:
false, timeouts: {implicit: 0, pageLoad: 300000, script: 30000},
unhandledPromptBehavior: dismiss and notify}
Session ID: d4843f8901ad0fe8956618e472c8dd1e
*** Element info: {Using=id, value=my-button5}
    at sun.reflect.NativeConstructorAccessorImpl.newInstance0(Native Method)
    at sun.reflect.NativeConstructorAccessorImpl.
newInstance(NativeConstructorAccessorImpl.java:62)
    at sun.reflect.DelegatingConstructorAccessorImpl.
newInstance(DelegatingConstructorAccessorImpl.java:45)
    at java.lang.reflect.Constructor.newInstance(Constructor.java:423)
    at org.openqa.selenium.remote.http.W3CHttpResponseCodec.
createException(W3CHttpResponseCodec.java:187)
    at org.openqa.selenium.remote.http.W3CHttpResponseCodec.
decode(W3CHttpResponseCodec.java:122)
    at org.openqa.selenium.remote.http.W3CHttpResponseCodec.
decode(W3CHttpResponseCodec.java:49)
    at org.openqa.selenium.remote.HttpCommandExecutor.
execute(HttpCommandExecutor.java:158)
    at org.openqa.selenium.remote.service.DriverCommandExecutor.
execute(DriverCommandExecutor.java:83)
    at org.openqa.selenium.remote.RemoteWebDriver.
execute(RemoteWebDriver.java:552)
    at org.openqa.selenium.remote.RemoteWebDriver.
findElement(RemoteWebDriver.java:323)
    at org.openqa.selenium.remote.RemoteWebDriver.
findElementById(RemoteWebDriver.java:372)
    at org.openqa.selenium.By$ById.findElement(By.java:188)
    at org.openqa.selenium.remote.RemoteWebDriver.
findElement(RemoteWebDriver.java:315)
    at com.lujiatao.webapplicationtest.SeleniumTest.main(SeleniumTest.java:21)
等待了 10 秒!
```

下面对运行结果进行说明。

① 使用 implicitlyWait()方法设置全局等待时间最长为 10s。

② 因为 ID 为“my-button3”的元素存在，且在打开 index-demo 页面时即被加载，所以等待时间为 0s。ID 为“my-button5”的元素不存在，因此在等待 10s 后抛出了 NoSuchElementException（没有该元素）异常。

③ 最后调用浏览器驱动的 quit()方法关闭浏览器。

6.4　TestNG 集成 Selenium

在待测程序中有一个 index 页面，该页面与 index-demo 页面相似，只是页面右上角多了一个登录用户的用户名和注销链接，如图 6-11 所示。

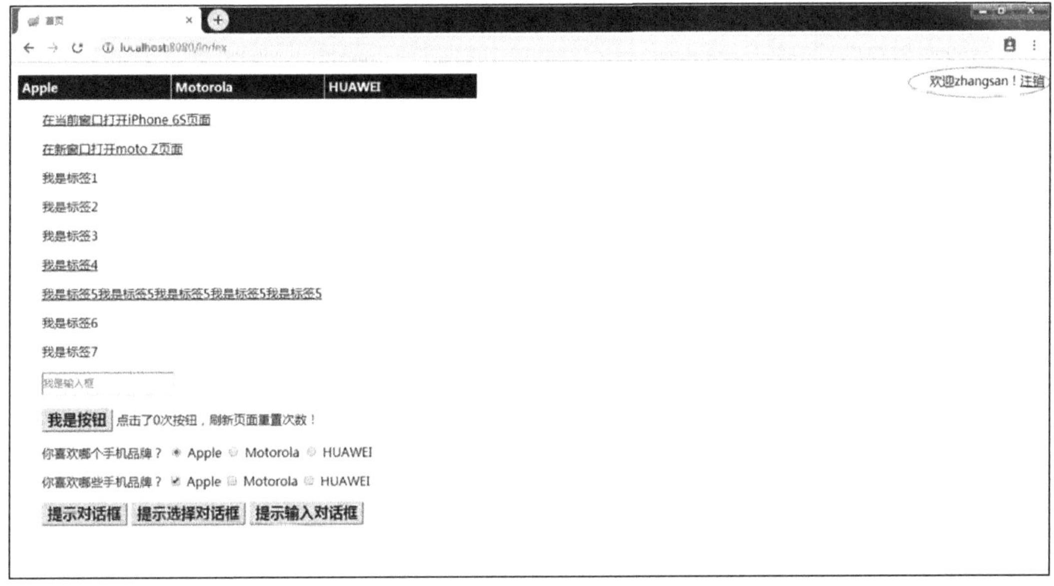

图 6-11

进入该页面需要登录，登录页面如图 6-12 所示。

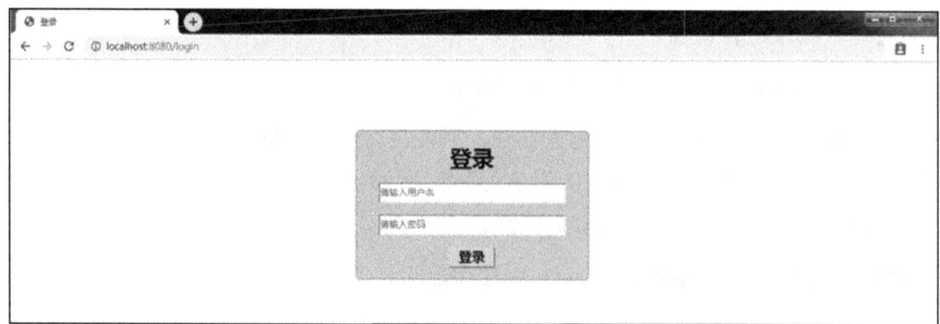

图 6-12

本节使用 TestNG 集成 Selenium 演示一个完整的自动化测试用例，该用例的功能为登录后判断是否登录成功。成功的依据为页面右上角显示登录的用户名与登录页面输入的用户名一致。

在 pom.xml 文件的<dependencies>标签中输入以下粗体部分的内容。

```
<project xmlns="http://maven.apache……"
    xmlns:xsi="http://www.w3……"
    xsi:schemaLocation="http://maven.apache…… http://maven.apache……">
    <modelVersion>4.0.0</modelVersion>

    <groupId>com.lujiatao</groupId>
    <artifactId>webapplicationtest</artifactId>
    <version>0.0.1-SNAPSHOT</version>
    <name>Web Application Test</name>

    <dependencies>
        <dependency>
            <groupId>org.seleniumhq.selenium</groupId>
            <artifactId>selenium-java</artifactId>
            <version>3.141.59</version>
        </dependency>
        <dependency>
            <groupId>org.testng</groupId>
            <artifactId>testng</artifactId>
            <version>6.14.3</version>
            <scope>test</scope>
        </dependency>
    </dependencies>

</project>
```

保存 pom.xml 文件，这时 Maven 会自动下载 TestNG 及其依赖的其他 jar 包。

在依赖 jar 包下载完成后，在工程（webapplicationtest）上单击右键，从弹出的快捷菜单中选择“TestNG → Convert to TestNG”选项，在工程中生成 testng.xml 文件。

删除 SeleniumTest 中的内容，输入以下代码。

```
package com.lujiatao.webapplicationtest;

import org.openqa.selenium.By;
import org.openqa.selenium.WebDriver;
import org.openqa.selenium.chrome.ChromeDriver;
import org.testng.Assert;
import org.testng.annotations.AfterClass;
import org.testng.annotations.BeforeClass;
import org.testng.annotations.Test;

public class SeleniumTest {
    private WebDriver chromeDriver;

    @BeforeClass
    public void init() {
        chromeDriver = new ChromeDriver();
    }

    @Test
    public void testCase1() {
        chromeDriver.get("http://localhost:8080/login");
        chromeDriver.findElement(By.name("username")).sendKeys("zhangsan");
        chromeDriver.findElement(By.name("password")).
            sendKeys("zhangsan123456");
        chromeDriver.findElement(By.xpath("//form/p[4]/button")).click();
        String expected = "zhangsan";
        String actual = chromeDriver.findElement(By.xpath("//nav/label/span")).
            getText();
        Assert.assertEquals(expected, actual);
    }

    @AfterClass
    public void clear() {
        chromeDriver.quit();
    }

}
```

修改 testng.xml 文件，在<test>标签中新增以下粗体部分内容。

```
<?xml version="1.0" encoding="UTF-8"?>
<!DOCTYPE suite SYSTEM "http://testng.org/testng-1.0.dtd">
<suite name="Suite">
    <test thread-count="5" name="Test">
        <classes>
            <class name="com.lujiatao.webapplicationtest.SeleniumTest" />
        </classes>
    </test> <!-- Test -->
</suite> <!-- Suite -->
```

保存所做的修改，在 testng.xml 上用鼠标右击，从弹出的快捷菜单中选择“Run As → TestNG

Suite”选项，查看测试报告，如图 6-13 所示。

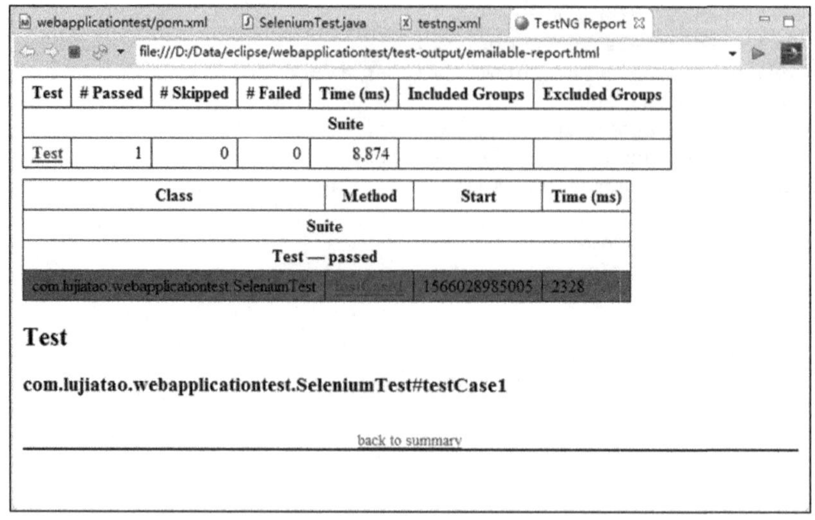

Test	# Passed	# Skipped	# Failed	Time (ms)	Included Groups	Excluded Groups
Suite						
Test	1	0	0	8,874		

Class	Method	Start	Time (ms)
Suite			
Test — passed			
com.lujiatao.webapplicationtest.SeleniumTest	testCase1	1566028985005	2328

Test

com.lujiatao.webapplicationtest.SeleniumTest#testCase1

back to summary

图 6-13

下面对运行结果进行说明。

① init()方法初始化了一个 Chrome 浏览器驱动，并在测试用例方法（testCase1）中使用，最后在 clear()方法中关闭。

② 通过 Name 定位用户名和密码输入框，然后输入用户名和密码。

③ 通过 XPath 定位登录按钮并单击，在 index 页面通过 XPath 定位右上角的用户名并获取文本。

④ 使用 TestNG 的断言方法对预期和实际结果进行比较，完成断言。

第 7 章 Android 自动化测试

7.1 Android 自动化测试工具（框架）简介

Android 操作系统基于 Linux 内核，应用的开发语言为 Kotlin（推荐）或 Java。Android 的自动化测试工具（框架）除 UFT、TestComplete 和 Katalon Studio 外，常见的还有以下 3 个。

① Appium：Appium 主要用于 Android 和 iOS 操作系统的原生、混合和 Web 应用（支持 iOS 上的 Safari，以及 Android 上的 Chrome 或自带浏览器）的自动化测试，另外，还支持 Windows 应用的自动化测试。自动化测试用例语言支持 Java、Python、JavaScript、Ruby、Objective-C、PHP、C#和 Perl 等。

② UI Automator：Google 官方 UI 自动化测试框架。UI Automator 仅支持 Android 4.3（API level 18）以上的 Android 版本，该框架支持跨应用。另外，可以使用 uiautomatorviewer 工具通过界面的方式扫描和分析当前 Android 设备上显示的 UI 组件。

③ Espresso：Google 官方 UI 自动化测试框架。与 UI Automator 不同的是，Espresso 仅支持单个应用的自动化测试，不支持跨应用。Espresso 可与 Android 单元测试框架 AndroidJUnitRunner 配合使用。Espresso 对 Android 版本的支持范围大于 UI Automator，可支持 Android 2.3.3（API level 10）以上的 Android 版本。

本章使用 Appium 作为 Android 自动化测试的工具（框架），并通过集成到 TestNG 来提高自动化测试用例的执行和管理效率。

7.2 安装待测应用

从笔者的 GitHub 主页下载待测应用，应用名为 Calculator.For.PPI.apk。下载后将应用拷贝至手机安装，安装完成后打开应用，如图 7-1 所示。

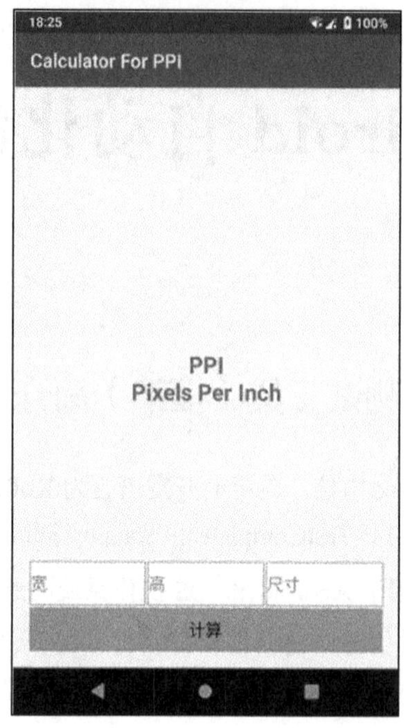

图 7-1

7.3 Appium 用法

7.3.1 准备

1．安装 Android Studio

具体安装方法，本书不再赘述，有兴趣的读者可到博文官网下载查看

2．安装 Appium

Appium 服务器在接收到自动化测试用例的指令后，会将指令发送给手机进行执行。Appium 服务器的作用类似于 Web 自动化测试中的浏览器驱动。

Appium 的下载地址见官网。下载后双击 Appium-windows-1.13.0.exe 文件，打开安装向导，勾选"为使用这台电脑的任何人安装（所有用户）"，单击"安装"按钮开始安装 Appium。在安装完成后单击“完成”按钮进入 Appium 首页。

单击“Start Server v1.13.0”按钮启动 Appium 服务器。

3. 安装 Android 手机驱动

针对不同的手机品牌和型号，安装的驱动会有所差异。将手机通过数据线连接到电脑后，建议登录对应手机厂商的官方网站下载并安装驱动，当然也可以使用应用宝等直接安装驱动。

当驱动安装成功后，打开手机的 USB 调试功能，此时通过 adb 命令可以查看到手机，如图 7-2 所示。

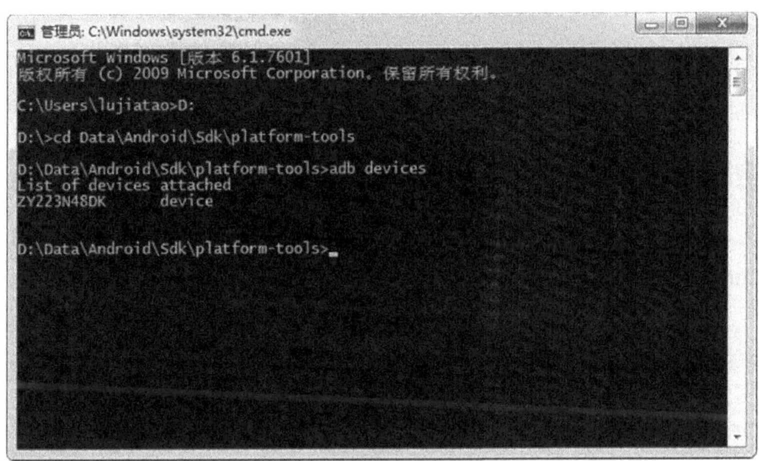

图 7-2

图 7-2 中的“ZY223N48DK”为设备名称。

4. 创建工程

创建一个新的 Maven 项目，关键信息填写如下。

Group Id：com.lujiatao

Artifact Id：androidapplicationtest

Name：Android Application Test

读者可根据实际情况填写，不需要和笔者填写的完全一致。

创建完成后，在 pom.xml 文件的<name>标签后输入以下粗体部分内容：

```
<project xmlns="http://maven.apache……"
    xmlns:xsi="http://www.w3……"
    xsi:schemaLocation="http://maven.apache…… http://maven.apache……">
    <modelVersion>4.0.0</modelVersion>
```

```
    <groupId>com.lujiatao</groupId>
    <artifactId>androidapplicationtest</artifactId>
    <version>0.0.1-SNAPSHOT</version>
    <name>Android Application Test</name>

    <dependencies>
        <dependency>
            <groupId>io.appium</groupId>
            <artifactId>java-client</artifactId>
            <version>7.1.0</version>
        </dependency>
    </dependencies>

</project>
```

保存 pom.xml 文件，这时 Maven 会自动下载 Appium 客户端及其依赖的其他 jar 包。

7.3.2 初始化参数

Appium 在启动手机 App 时需要传入一些初始化参数，这些初始化参数是通过 DesiredCapabilities 对象来指定的。常用的初始化参数如表 7-1 所示。

表 7-1

参数名	描　述	支持系统
automationName	自动化测试引擎，Appium（默认）或Selendroid	Android&iOS
platformName	操作系统，Android或iOS	Android&iOS
platformVersion	操作系统版本	Android&iOS
deviceName	设备名称，Android Emulator、Galaxy S4和iPhone Simulator等	Android&iOS
app	安装包路径，Android如果指定了appPackage和appActivity，则不需要该参数	Android&iOS
browserName	使用的浏览器名称，Safari、Chrome等	Android&iOS
newCommandTimeout	Appium服务器等待超时时间	Android&iOS
language	设置模拟器语言	Android&iOS
locale	设置模拟器区域	Android&iOS
orientation	设置模拟器方向，竖屏或横屏	Android&iOS
autoWebview	自动转换到Webview上下文，false（默认）或true。	Android&iOS
noReset	不重置应用，false（默认）或true	Android&iOS
fullReset	清除应用，测试完成会将应用卸载掉，false（默认）或true	Android&iOS
appActivity	Android的启动Activity	Android

续表

参数名	描　述	支持系统
appPackage	Android应用包名	Android
appWaitDuration	应用启动超时时间，默认为20000（毫秒）	Android
deviceReadyTimeout	设备准备就绪超时时间	Android
androidDeviceReadyTimeout	启动应用后准备就绪超时时间	Android
androidInstallTimeout	安装应用超时时间	Android
avd	模拟器名称	Android
avdLaunchTimeout	模拟器启动超时时间	Android
avdReadyTimeout	模拟器准备就绪超时时间	Android
avdArgs	模拟器额外参数	Android
bundleId	应用的bundle ID	iOS
launchTimeout	Appium服务器等待超时时间	iOS
locationServicesEnabled	模拟器强制打开或关闭定位服务	iOS
interKeyDelay	按键之间的延迟时间，以毫秒为单位	iOS
appName	应用名称	iOS

初始化参数分为通用类和专用类两种，通用类适用于所有平台，专用类则只支持 Android 或 iOS。

在介绍使用初始化参数启动 Android 应用之前，先获取两个启动参数的值，即 appPackage 和 appActivity。在手机上打开已安装的待测 Android 应用，再打开 CMD 窗口，执行以下命令。

```
adb shell dumpsys window | findstr mCurrentFocus
```

执行后可以看到 appPackage 和 appActivity，如图 7-3 所示。

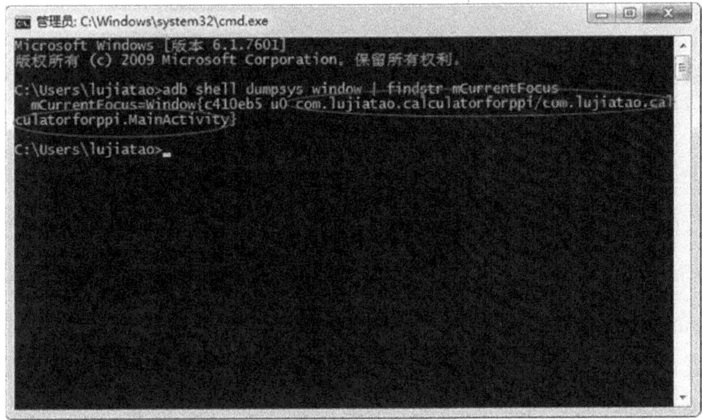

图 7-3

接下来在 src/test/java 中创建名为 com.lujiatao.androidapplicationtest 的 Package 及名为 AndroidTest 的 Class（需要勾选“public static void main(String[] args)”），在 AndroidTest 中输入以下代码。

```
package com.lujiatao.androidapplicationtest;

import java.net.MalformedURLException;
import java.net.URL;

import org.openqa.selenium.remote.DesiredCapabilities;

import io.appium.java_client.android.AndroidDriver;
import io.appium.java_client.android.AndroidElement;

public class AndroidTest {

    public static void main(String[] args) {
        AndroidDriver<AndroidElement> driver = null;
        DesiredCapabilities capabilities = new DesiredCapabilities();
        capabilities.setCapability("platformName", "Android");
        capabilities.setCapability("platformVersion", "9");
        capabilities.setCapability("deviceName", "My Android Device");
        capabilities.setCapability("appPackage", "com.lujiatao.calculatorforppi");
        capabilities.setCapability("appActivity",
            "com.lujiatao.calculatorforppi.MainActivity");
        capabilities.setCapability("noReset", "true");
        try {
            driver = new AndroidDriver<AndroidElement>(new
                URL("http://localhost:4723/wd/hub"), capabilities);
        } catch (MalformedURLException e) {
            e.printStackTrace();
        }
    }

}
```

保存代码，按快捷键 F11 运行工程，此时连接电脑的手机会自动启动待测应用并进入应用首页，如图 7-1 所示。

需要说明的是，在运行上述代码之前需保证手机屏幕已解锁。

现对以上运行结果进行说明。

① 启动参数的 platformName 设置为 Android。platformVersion 可以通过查看手机版本号获得。deviceName 的值目前在 Android 中已被忽略，但该参数仍然需要设置。appPackage 和 appActivity 已通过命令行获得。noReset 设置为 true，避免重置应用，尤其是有欢迎页面或登录页面的应用在重置后，会导致后续启动应用不能直接进入应用主页。

② 构造一个 Android 驱动，参数有两个。一个是 URL 对象，该对象描述了 Appium 服务器的地址，通过与服务器建立 HTTP 会话来传输数据，这一点从图 7-4 可以看出；另一个是初始化参数对象 DesiredCapabilities。

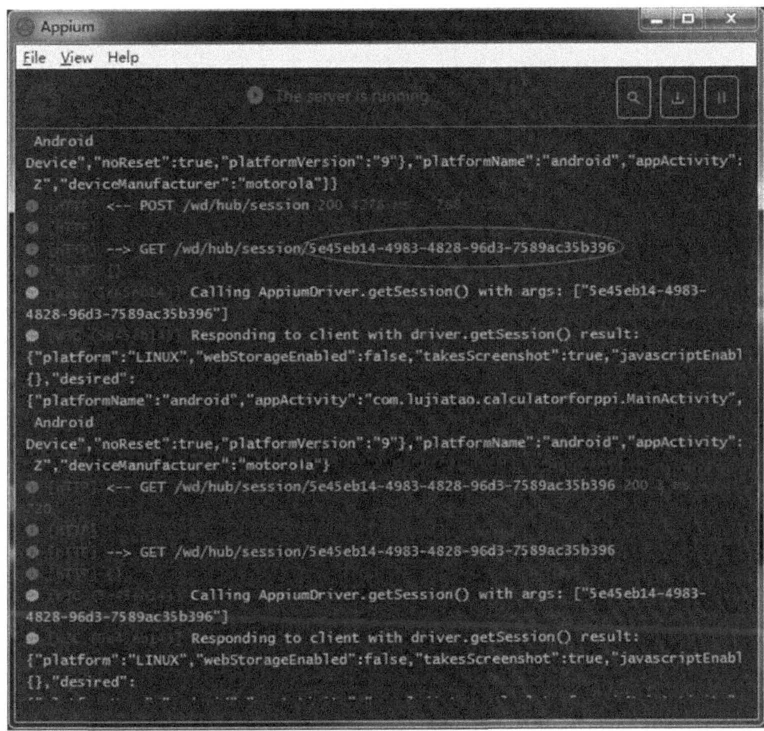

图 7-4

7.3.3　元素操作

1. 元素查看

在 Web 自动化测试中，可以通过浏览器的开发者工具直接查看页面元素，但在 Android 手机上并不能直接查看元素，而是需要通过 Android SDK 中的 uiautomatorviewer 工具或 Appium 查看，Appium 的相关内容会在第 9 章介绍，本章仅介绍 uiautomatorviewer 的查看方法。

uiautomatorviewer 在 Android SDK 的 tools\bin 目录下，双击 uiautomatorviewer.bat 即可打开。此时保持手机连接电脑，打开待测应用，单击“Device Screenshot”图标，如图 7-5 所示。

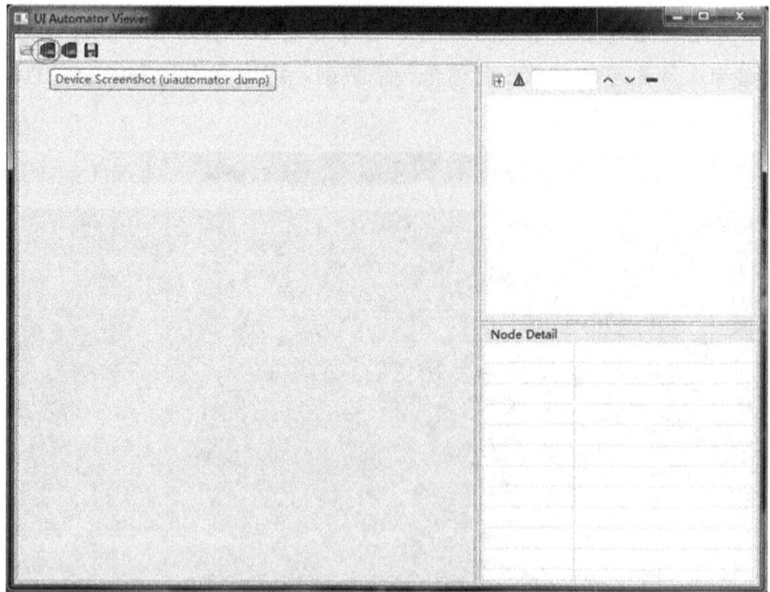

图 7-5

稍等片刻，待测应用的页面元素便显示了出来。

部分手机会提示“Error taking device screenshot: EOF”（无法查看元素），这种情况可以通过手动执行命令实现。

① 打开 CMD 窗口，依次执行以下命令，如图 7-6 所示。

```
adb shell uiautomator dump /sdcard/myapp.uix
adb shell screencap -p /sdcard/app.png
adb pull /sdcard/myapp.uix
adb pull /sdcard/app.png
```

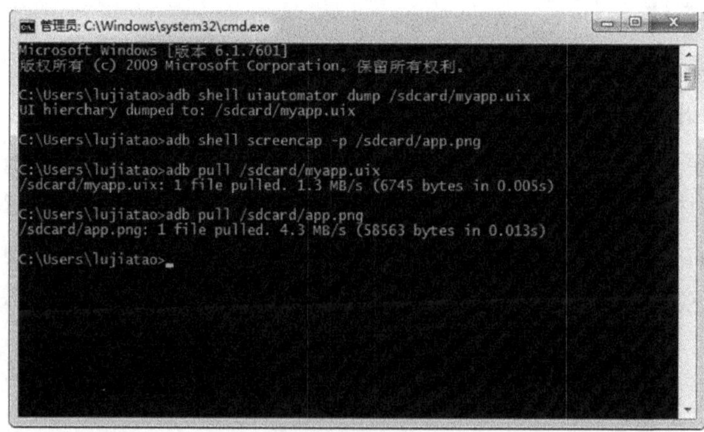

图 7-6

执行以上命令后，会将当前屏幕的布局和截图上传到本地电脑的 C:\Users\lujiatao 目录，命令中的 myapp.uix 和 app.png 可根据实际情况随意命名。

② 单击 uiautomatorviewer 的“Open”图标，打开上一步获取的两个文件，此时元素显示在了 uiautomatorviewer 的右侧，如图 7-7 所示。

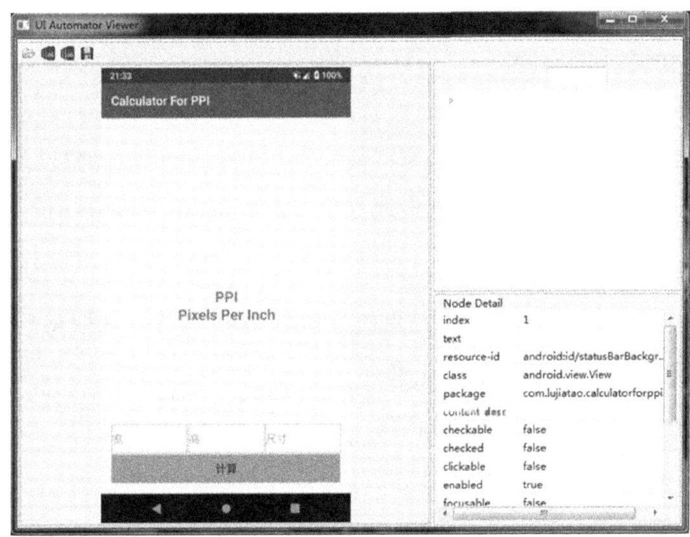

图 7-7

将鼠标光标移动到左侧元素上，右侧就会显示对应元素的相关信息。

2. 定位元素

与 Selenium 类似，在 Appium 中也有 8 种定位元素的方法。

- findElementByClassName：通过 Class 定位元素。
- findElementByCssSelector：通过 CSS 选择器定位元素。
- findElementById：通过 ID 定位元素。
- findElementByLinkText：通过超链接文本定位元素。
- findElementByName：通过 Name 定位元素。
- findElementByPartialLinkText：通过超链接的部分文本定位元素。
- findElementByTagName：通过标签名定位元素。
- findElementByXPath：通过 XPath 定位元素。

以上方法只适用于返回一个元素的情况，想要返回多个元素，则可以使用对应的 findElementsByXXX 方法，如下所示。

- findElementsByClassName

- findElementsByCssSelector
- findElementsById
- findElementsByLinkText
- findElementsByName
- findElementsByPartialLinkText
- findElementsByTagName
- findElementsByXPath

另外，可以通过传入 By 对象获取元素，对应的方法为 findElement()和 findElements()。

除上述方法外，Android 中还有一些专属方法，比如 findElementByAndroidUIAutomator()。

3. 操作元素

对元素的操作也与 Selenium 类似，常见的有单击、输入文本、获取文本和获取属性等。

下面举例说明如何定位元素和操作元素。

在 AndroidTest 中输入以下粗体部分代码。

```
package com.lujiatao.androidapplicationtest;

import java.net.MalformedURLException;
import java.net.URL;

import org.openqa.selenium.remote.DesiredCapabilities;

import io.appium.java_client.android.AndroidDriver;
import io.appium.java_client.android.AndroidElement;

public class AndroidTest {

    public static void main(String[] args) {
        AndroidDriver<AndroidElement> driver = null;
        DesiredCapabilities capabilities = new DesiredCapabilities();
        capabilities.setCapability("platformName", "Android");
        capabilities.setCapability("platformVersion", "9");
        capabilities.setCapability("deviceName", "My Android Device");
        capabilities.setCapability("appPackage", "com.lujiatao.calculatorforppi");
        capabilities.setCapability("appActivity",
            "com.lujiatao.calculatorforppi.MainActivity");
        capabilities.setCapability("noReset", "true");
        try {
            driver = new AndroidDriver<AndroidElement>(new
                URL("http://localhost:4723/wd/hub"), capabilities);
        } catch (MalformedURLException e) {
            e.printStackTrace();
        }
```

```
        driver.findElementById("com.lujiatao.calculatorforppi:id/editText1").
            sendKeys("750");
        driver.findElementById("com.lujiatao.calculatorforppi:id/editText2").
            sendKeys("1334");
        driver.findElementById("com.lujiatao.calculatorforppi:id/editText3").
            sendKeys("4.7");
        driver.findElementByClassName("android.widget.Button").click();
        String result = driver.findElementByXPath(
        "//android.view.ViewGroup/android.widget.FrameLayout[2]/android.
             view.ViewGroup/android.widget.TextView").getText();
        System.out.println(result);
    }

}
```

保存代码，按快捷键 F11 运行工程， Eclipse 的控制台输出如下：

```
八月 22, 2019 10:11:38 下午 io.appium.java_client.remote.AppiumCommandExecutor$1
lambda$0
信息: Detected dialect: W3C
326 ppi
```

此时待测应用会自动打开，并显示如图 7-8 所示的内容。

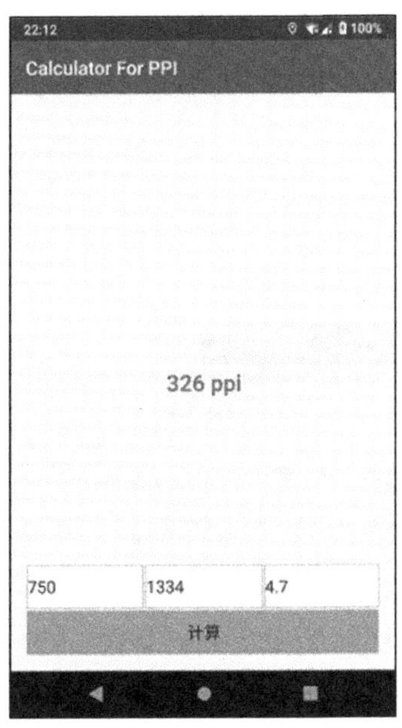

图 7-8

下面对运行结果进行说明。

① 在 uiautomatorviewer 中 Node Detail 显示的 resource-id 值即元素的 ID，这里用 ID 定位了 3 个输入框并输入了文本。

② 使用 Class 定位并单击“计算”按钮，Class 取值为 uiautomatorviewer 中 Node Detail 显示的 class 值。

③ 使用 XPath 定位 TextView（Android 中用于显示文本的组件），XPath 路径的获取方法如下。

- 使用 uiautomatorviewer 定位要获取 XPath 的元素，如图 7-9 所示。

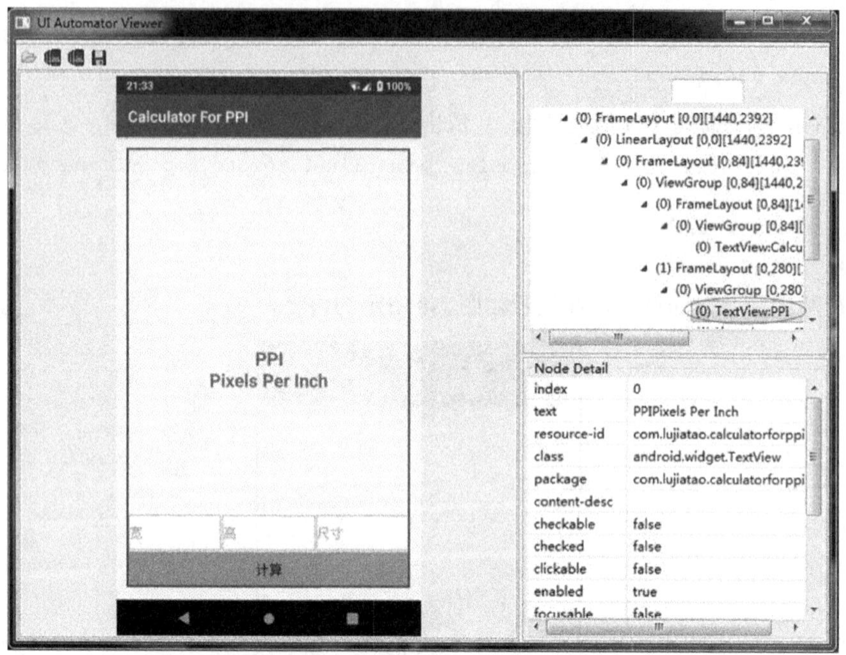

图 7-9

- 观察层级关系：父元素的父元素（FrameLayout）是并列的，且下面均有 ViewGroup 和 TextView，因此想要区分该 TextView，就需要从 FrameLayout 的父元素开始往下找。最终确定 XPath 的写法为“//android.view.ViewGroup/android.widget.FrameLayout[2]/android.view.ViewGroup/android.widget.TextView”。括号里的“2”代表第 2 个（index 为 1）。

④ 使用 getText()方法获取元素文本，并将文本打印到控制台，文本对应 uiautomatorviewer 中 Node Detail 显示的 text 值。

7.3.4　应用操作

本节介绍一些应用级别的操作，在进行这些操作之前，需要获取 Android 驱动对象。常用操作如下。

- isAppInstalled()：检测 Android 应用是否已安装，参数为应用的包名，比如“com.lujiatao.calculatorforppi”。
- installApp()：安装 Android 应用，参数为应用安装包的路径名，比如“E:\\Calculator.For.PPI.apk”。
- removeApp()：卸载 Android 应用，参数为应用的包名。
- activateApp()：启动 Android 应用。如果应用未运行，则打开应用；如果应用在后台，则将应用重新置于前台，参数为应用的包名。
- closeApp()：关闭 Android 应用，相当于按下 Home 键，只是将应用放到后台，并不是真正的关闭。

下面举例说明。

删除 AndroidTest 中的内容，输入以下代码：

```
package com.lujiatao.androidapplicationtest;

import java.net.MalformedURLException;
import java.net.URL;

import org.openqa.selenium.remote.DesiredCapabilities;

import io.appium.java_client.android.AndroidDriver;
import io.appium.java_client.android.AndroidElement;

public class AndroidTest {

    public static void main(String[] args) {
        AndroidDriver<AndroidElement> driver = null;
        DesiredCapabilities capabilities = new DesiredCapabilities();
        capabilities.setCapability("platformName", "Android");
        capabilities.setCapability("platformVersion", "9");
        capabilities.setCapability("deviceName", "My Android Device");
        capabilities.setCapability("appPackage", "com.lujiatao.calculatorforppi");
        capabilities.setCapability("appActivity",
            "com.lujiatao.calculatorforppi.MainActivity");
        capabilities.setCapability("noReset", "true");
        try {
            driver = new AndroidDriver<AndroidElement>(new
                URL("http://localhost:4723/wd/hub"), capabilities);
        } catch (MalformedURLException e) {
            e.printStackTrace();
        }
```

```
        boolean isInstalled = driver.isAppInstalled("com.android.calculator2");
        if (isInstalled) {
            driver.activateApp("com.android.calculator2");
        }
    }

}
```

保存代码，按快捷键 F11 运行工程，此时待测应用会自动打开，同时打开手机自带的计算器应用，如图 7-10 所示。

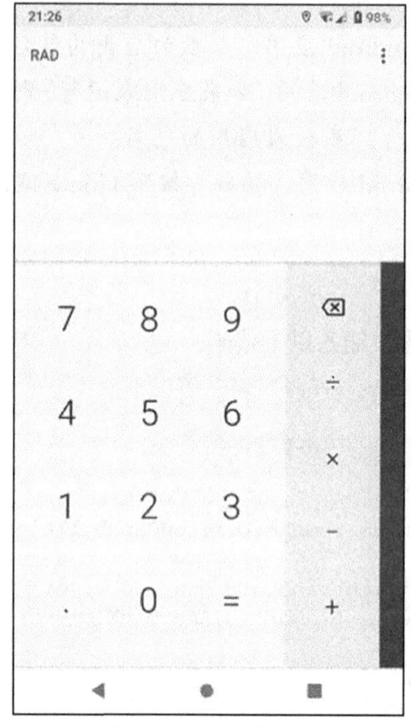

图 7-10

这里先判断是否安装了计算器应用。如果已安装，则启动应用。至于如何查看 Android 应用的包名，本书前面已有介绍，这里不再赘述。

7.3.5 系统操作

系统操作是比应用操作更底层的一些操作，比如锁屏、解锁屏幕和模拟键盘输入等操作，常用操作如下。

- isDeviceLocked()：检测屏幕是否已锁屏。
- lockDevice()：锁屏。

- unlockDevice()：解锁屏幕。
- sendKeys()：模拟键盘输入文本。
- hideKeyboard()：隐藏系统键盘。

下面举例说明。

将 AndroidTest 的 main()方法中最后 4 行代码替换为以下代码。

```
boolean isLocked = driver.isDeviceLocked();
if (!isLocked) {
    driver.lockDevice();
}
```

保存代码，按快捷键 F11 运行工程，此时待测应用会自动打开，打开后自动锁屏。

7.3.6 使用 Android 模拟器

之前的 Android 应用测试都是在真实手机上进行的，在 Android 应用开发过程中，经常基于模拟器进行调试。当然 Android 应用测试也可以基于模拟器进行测试。在 Android Studio 首页单击“Configure → AVD Manager”进入模拟器列表，双击其中一个模拟器，即可启动该模拟器。

值得注意的是，如果同时连接了多个待测设备，比如同时连接了一个真实手机和一个模拟器，那么只会在其中一个设备上运行测试用例。可以使用 adb 命令查看连接的多个设备，如图 7-11 所示。

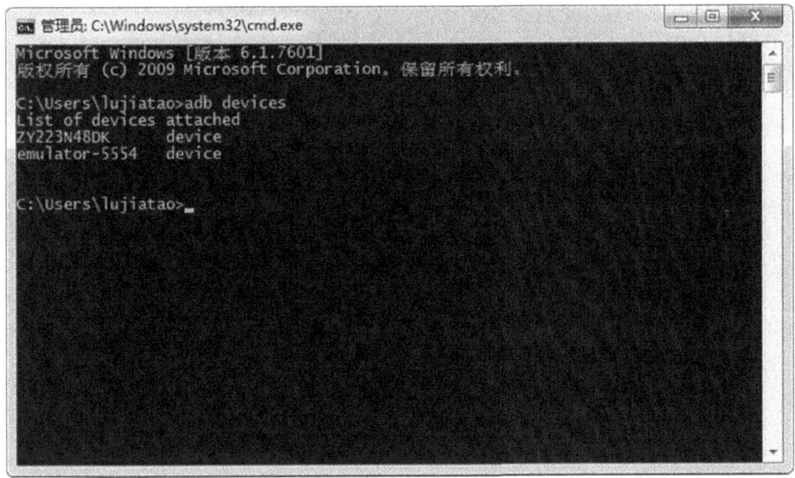

图 7-11

打开模拟器后，安装待测应用到模拟器，保持代码不变，按快捷键 F11 运行工程，此时待测应用会自动打开，当应用打开后会自动锁屏。

7.4 TestNG 集成 Appium

本节使用 TestNG 集成 Appium 演示一个完整的 Android 自动化测试用例。该用例的功能为启动待测应用，计算 iPhone 6S 的 PPI，并与预期值进行对比。

在 pom.xml 文件的<dependencies>标签中输入以下粗体部分内容：

```
<project xmlns="http://maven.apache:……"
    xmlns:xsi="http://www.w3……"
    xsi:schemaLocation="http://maven.apache…… http://maven.apache……">
    <modelVersion>4.0.0</modelVersion>

    <groupId>com.lujiatao</groupId>
    <artifactId>androidapplicationtest</artifactId>
    <version>0.0.1-SNAPSHOT</version>
    <name>Android Application Test</name>

    <dependencies>
        <dependency>
            <groupId>io.appium</groupId>
            <artifactId>java-client</artifactId>
            <version>7.1.0</version>
        </dependency>
        <dependency>
            <groupId>org.testng</groupId>
            <artifactId>testng</artifactId>
            <version>6.14.3</version>
            <scope>test</scope>
        </dependency>
    </dependencies>

</project>
```

保存 pom.xml 文件，这时 Maven 会自动下载 TestNG 及其依赖的其他 jar 包。

依赖 jar 包下载完成后，在工程（androidapplicationtest）上用鼠标右击，从弹出的快捷菜单中选择“TestNG → Convert to TestNG”选项，在工程中生成 testng.xml 文件。

删除 AndroidTest 中的内容，输入以下代码：

```
package com.lujiatao.androidapplicationtest;

import java.net.MalformedURLException;
import java.net.URL;

import org.openqa.selenium.remote.DesiredCapabilities;
import org.testng.Assert;
import org.testng.annotations.AfterClass;
import org.testng.annotations.BeforeClass;
import org.testng.annotations.Test;
```

```
import io.appium.java_client.android.AndroidDriver;
import io.appium.java_client.android.AndroidElement;

public class AndroidTest {
    private AndroidDriver<AndroidElement> driver = null;

    @BeforeClass
    public void init() {
        DesiredCapabilities capabilities = new DesiredCapabilities();
        capabilities.setCapability("platformName", "Android");
        capabilities.setCapability("platformVersion", "9");
        capabilities.setCapability("deviceName", "My Android Device");
        capabilities.setCapability("appPackage", "com.lujiatao.calculatorforppi");
        capabilities.setCapability("appActivity",
            "com.lujiatao.calculatorforppi.MainActivity");
        capabilities.setCapability("noReset", "true");
        try {
            driver = new AndroidDriver<AndroidElement>(new
                URL("http://localhost:4723/wd/hub"), capabilities);
        } catch (MalformedURLException e) {
            e.printStackTrace();
        }
    }

    @Test
    public void testCase1() {
        driver.findElementById("com.lujiatao.calculatorforppi:id/editText1").
            sendKeys("750");
        driver.findElementById("com.lujiatao.calculatorforppi:id/editText2").
            sendKeys("1334");
        driver.findElementById("com.lujiatao.calculatorforppi:id/editText3").
            sendKeys("4.7");
        driver.findElementByClassName("android.widget.Button").click();
        String expected = "326 ppi";
        String actual = driver.findElementByXPath(
                "//android.view.ViewGroup/android.widget.FrameLayout[2]/
                   android.view.ViewGroup/android.widget.TextView").getText();
        Assert.assertEquals(expected, actual);
    }

    @AfterClass
    public void clear() {
        driver.quit();
    }

}
```

修改 testng.xml 文件，在<test>标签中新增以下粗体部分内容。

```
<?xml version="1.0" encoding="UTF-8"?>
<!DOCTYPE suite SYSTEM "http://testng.org/testng-1.0.dtd">
```

```
<suite name="Suite">
    <test thread-count="5" name="Test">
        <classes>
            <class name="com.lujiatao.androidapplicationtest.AndroidTest" />
        </classes>
    </test> <!-- Test -->
</suite> <!-- Suite -->
```

保存所做的修改，在 testng.xml 上用鼠标右击，从弹出的快捷菜单中选择“Run As → TestNG Suite”选项，查看测试报告，如图 7-12 所示。

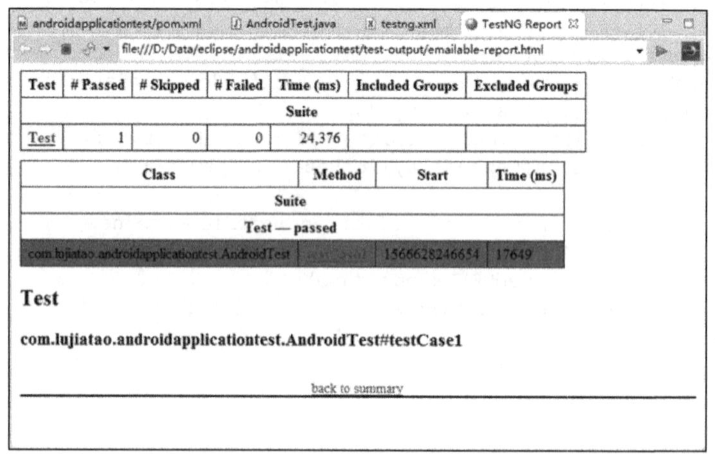

图 7-12

下面对运行结果进行说明。

① init()方法初始化了一个 Android 驱动对象，并在测试用例方法（testCase1）中使用，最后在 clear()方法中关闭。

② 通过 ID、Class 和 XPath 等方法定位和操作元素。

③ 使用 TestNG 的断言方法对预期和实际结果进行比较，完成断言。

第8章 iOS 自动化测试

8.1 iOS 自动化测试工具（框架）简介

iOS 操作系统基于 XNU 内核，应用的开发语言为 Swift（推荐）或 Objective-C。iOS 的自动化测试工具（框架）除了 UFT、TestComplete、Katalon Studio 和 Appium 外，常见的还有以下两个苹果公司官方推出的自动化测试框架。

UIAutomation：UIAutomation 随着 iOS 4.0 发布，支持手机和模拟器的 UI 自动化测试。在 iOS 9.3 之前，Appium 集成了 UIAutomation 进行 iOS 的自动化测试。

XCTest：XCTest 支持 iOS 的单元测试、UI 测试和性能测试，需要 Xcode 5.0 及以上版本的支持。XCTest 可以看成是 UIAutomation 的替代品，但功能更为强大。在 iOS 9.3 之后，Appium 集成了 XCTest 中的 XCUITest 进行 iOS 的自动化测试。

本章使用 Appium 作为 iOS 自动化测试的工具（框架），并通过集成到 TestNG 来提高自动化测试用例的执行和管理效率。

8.2 待测应用开发

苹果公司对应用的管理非常严格，在手机（非模拟器）上进行 iOS 自动化测试时需要对待测应用进行签名；并且在应用上架 App Store 之前，需采用 TestFlight 或第三方分发平台进行大规模的应用分发，但以上方式均有数量的限制，且第三方分发平台是付费的。基于上述原因，直接开发一个简单的待测应用不失为一个最佳选择，在开发时使用自己的证书进行签名即可。

由于 Swift 语法和 Java 语法差异较大，因此读者可直接按照示例编写代码，使其满足本章的自动化测试学习需求即可，不必深入理解其含义。

8.2.1 工程创建

在 Xcode 中选择“File → New → Project...”选项，进入配置模板页面。

默认为“Single View App”，单击“Next”按钮进入配置选项页面，如图 8-1 所示。

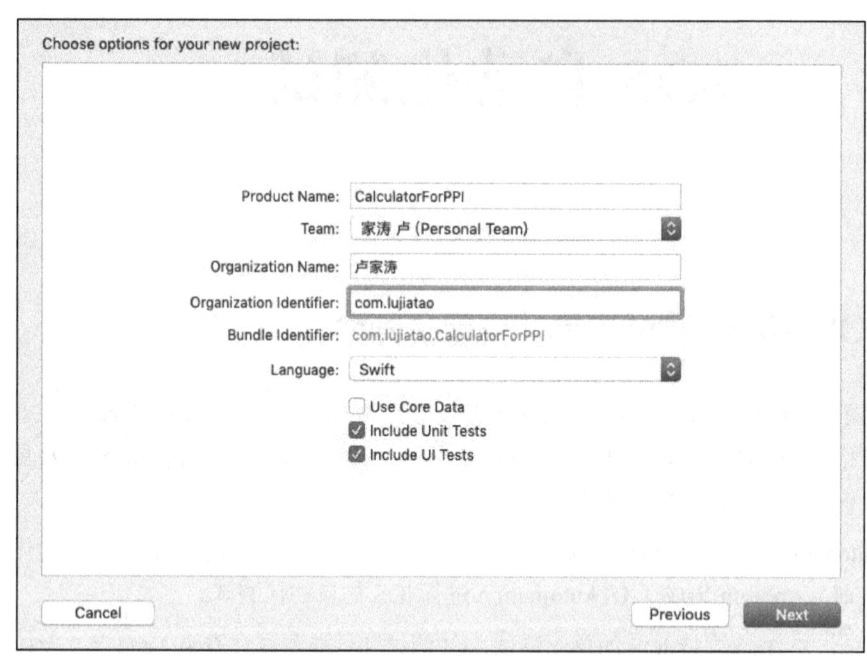

图 8-1

Product Name 和 Organization Name 需要手动填写，Team 通过下拉框选择即可，其他项保持不变。单击“Next”按钮，选择项目保存的位置即可创建工程。

8.2.2 界面开发

单击 Xcode 左侧的 Main.storyboard 文件，切换到 UI 设计页面，单击右上角的“Library”图标，打开对象库，如图 8-2 所示。

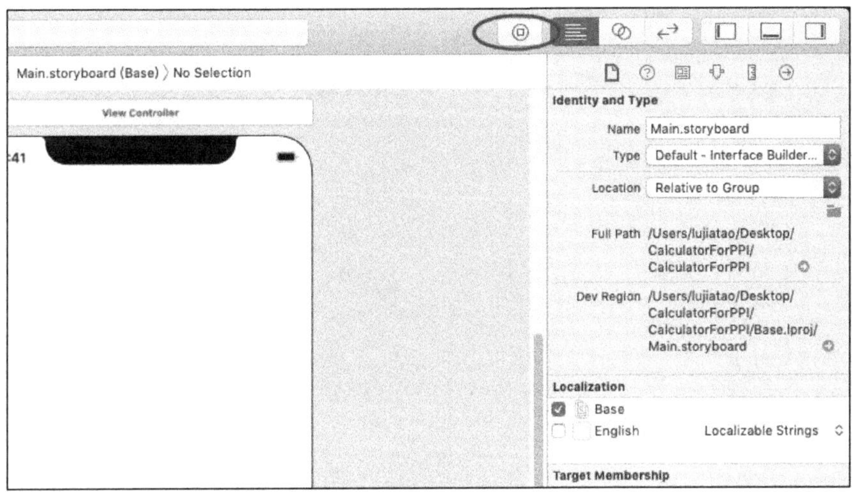

图 8-2

在对象库中，分别选择 1 个 Label、3 个 Text Field 和 1 个 Button，把它们拖放到 UI 设计页面。通过鼠标调整这些控件的位置和尺寸，修改 Label 中的文字内容、颜色和字号（设置为加粗（Bold）），修改 Label 对齐方式为居中，修改 Label 行数为 2，如图 8-3 所示。

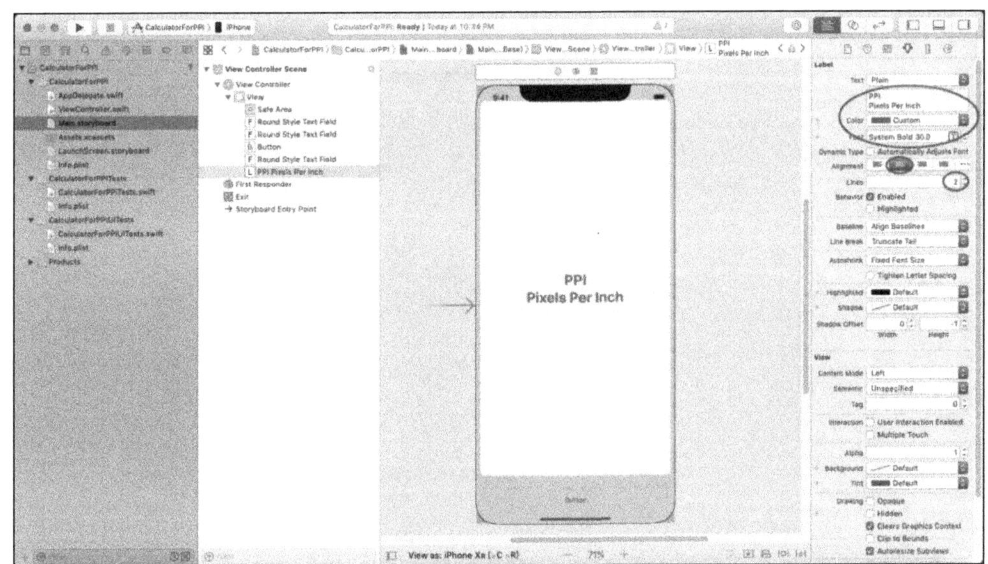

图 8-3

修改 3 个 Text Field 的默认文本，分别修改为“高”、“宽”和“尺寸”，如图 8-4 所示。

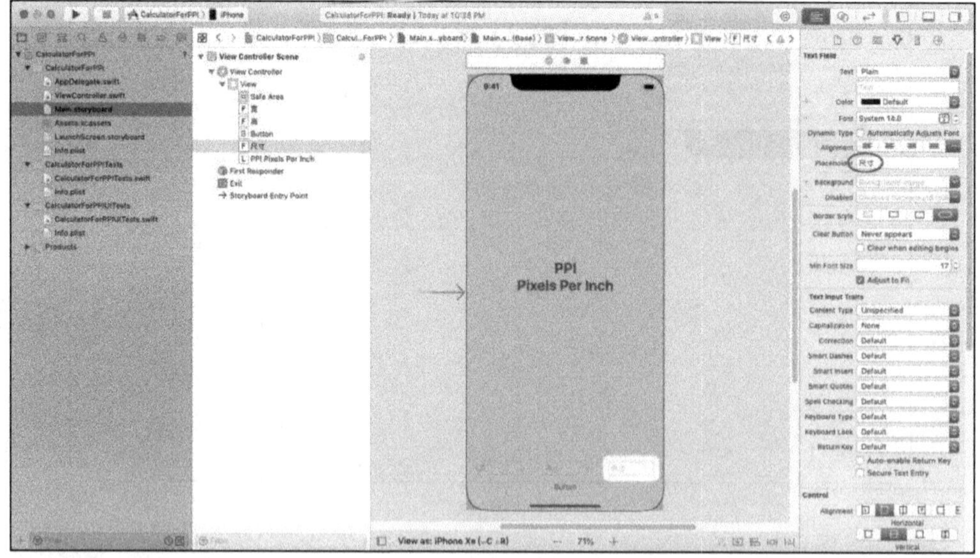

图 8-4

修改 Button 的文字内容、文字颜色和背景色，如图 8-5 所示。

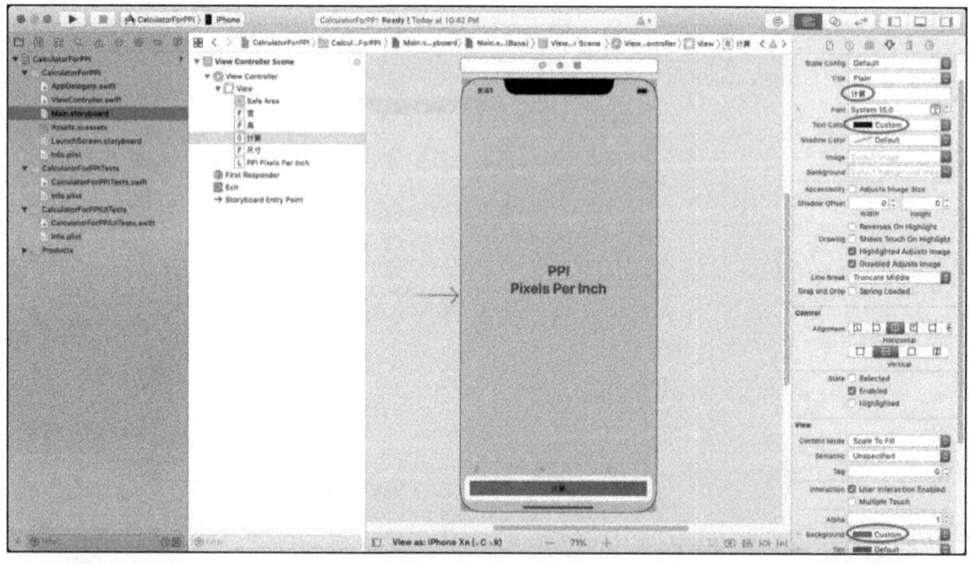

图 8-5

查看手机屏幕的适配情况，以上为 iPhone XR 的适配。由于笔者使用的是 iPhone 6S，屏幕分辨率和 iPhone 8 一致，因此需要将 Xcode 下方的“View as”修改为 iPhone 8，再拖动鼠标调整界面，调整完成后如图 8-6 所示。

图 8-6

8.2.3　逻辑开发

在进行逻辑开发之前，应先安装一个第三方框架 IQKeyboardManager，该框架用于处理虚拟键盘。这里使用 carthage 管理第三方包。首先执行以下命令，创建并打开 Cartfile：

```
cd /Users/lujiatao/Desktop/CalculatorForPPI/
touch Cartfile
open -a Xcode Cartfile
```

/Users/lujiatao/Desktop/为笔者放置工程的目录，读者可根据实际情况进行替换。打开后在 Cartfile 中加入 github "hackiftekhar/IQKeyboardManager"并保存，然后继续执行以下命令，即可安装 IQKeyboardManager：

```
carthage update --platform iOS
```

将 /Users/lujiatao/Desktop/CalculatorForPPI/Carthage/Build/iOS/IQKeyboardManagerSwift.framework 文件拖进 CalculatorForPPI 工程，然后在工程的 General 中加入该框架，如图 8-7 所示。

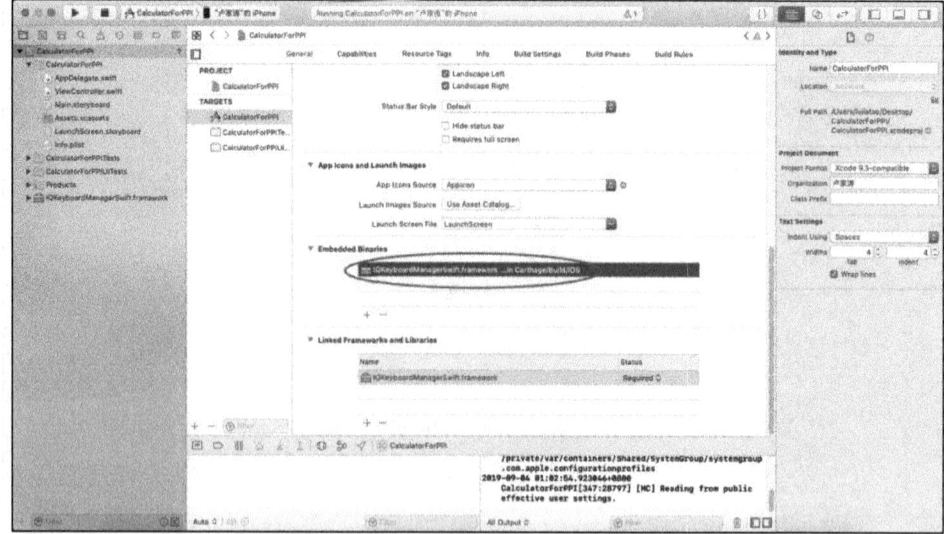

图 8-7

单击图 8-7 中右上角的两个圆圈图标打开“Assistant editor”，此时界面和代码可在同一个屏幕显示，如图 8-8 所示。

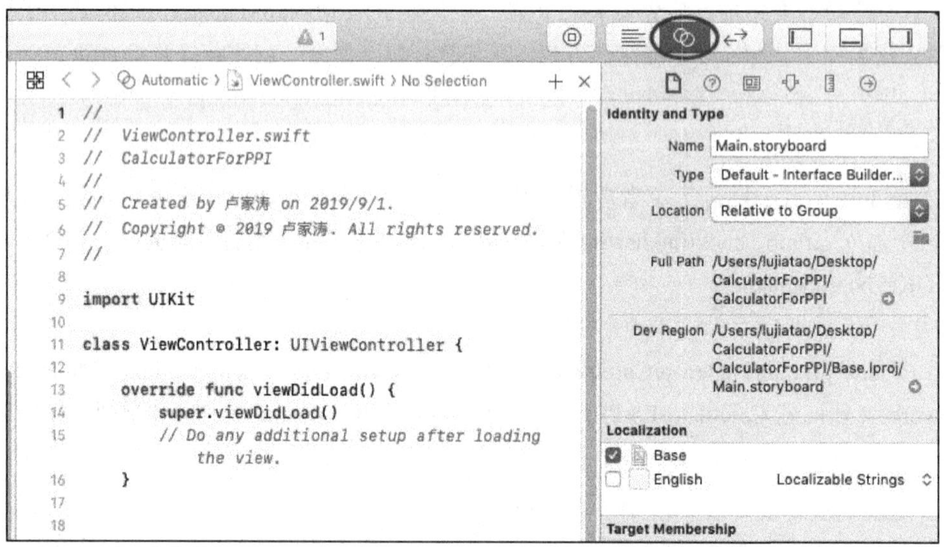

图 8-8

用鼠标选中左边的控件，按“control”键，将控件拖到右边代码中，分别将 1 个 Label、3 个 Text Field 和 1 个 Button 拖到右边定义成变量（Connection 选 Outlet），另外，再将 Button 拖到右边定义成方法（Connection 选 Action）。将 Label 定义成变量的示例如图 8-9 所示。

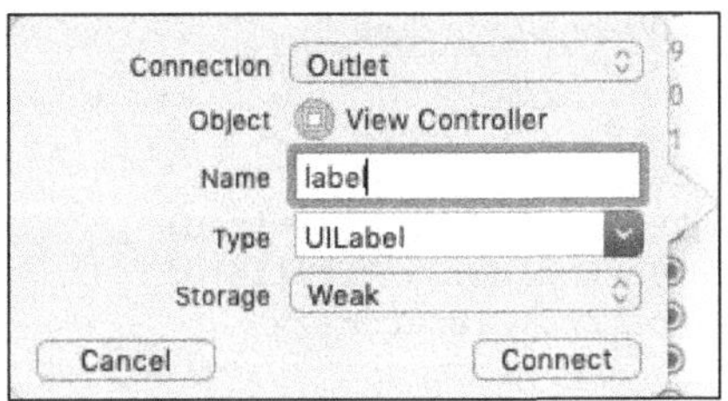

图 8-9

在 AppDelegate.swift 文件中导入 IQKeyboardManager 框架，并在 application()方法中启用该框架，如下粗体内容所示：

```
//
//  AppDelegate.swift
//  CalculatorForPPI
//
//  Created by 卢家涛 on 2019/9/1.
//  Copyright © 2019 卢家涛. All rights reserved.
//

import UIKit
import IQKeyboardManagerSwift

@UIApplicationMain
class AppDelegate: UIResponder, UIApplicationDelegate {

    var window: UIWindow?


    func application(_ application: UIApplication, didFinishLaunchingWithOptions
        launchOptions: [UIApplication.LaunchOptionsKey: Any]?) -> Bool {
        // Override point for customization after application launch.
        IQKeyboardManager.shared.enable = true
        return true
    }
// 余下代码省略
```

在 ViewController.swift 文件中编写如下代码：

```
//
//  ViewController.swift
//  CalculatorForPPI
//
//  Created by 卢家涛 on 2019/9/1.
//  Copyright © 2019 卢家涛. All rights reserved.
//

import UIKit
```

```
class ViewController: UIViewController {

    @IBOutlet weak var label: UILabel!
    @IBOutlet weak var textField1: UITextField!
    @IBOutlet weak var textField2: UITextField!
    @IBOutlet weak var textField3: UITextField!
    @IBOutlet weak var button: UIButton!

    override func viewDidLoad() {
        super.viewDidLoad()
    }

    @IBAction func calculate(_ sender: UIButton) {
        let width: Int = Int(textField1.text!)!
        let height: Int = Int(textField2.text!)!
        let size: Double = Double(textField3.text!)!
        let result: Int = Int(round(sqrt((pow(Double(width), 2) + pow(Double(height),
            2)) / pow(size, 2))))
        label.text = "\(result) ppi"
    }

    override func touchesBegan(_ touches: Set<UITouch>, with event: UIEvent?) {
        textField1.resignFirstResponder()
        textField2.resignFirstResponder()
        textField3.resignFirstResponder()
    }

}
```

现对上述代码进行说明。

① @IBOutlet 标记的变量是之前通过拖动创建的。

② @viewDidLoad()方法是重写了父类的 viewDidLoad()方法，在应用启动时调用。

③ @IBAction 标记的方法是之前通过拖动创建的，该方法首先获取 UITextField 中的值，然后计算 PPI，最后将结果显示在 UILabel 中。

④ TouchesBegan()方法是一个重写的方法，作用是监听单击事件，让 UITextField 失去焦点；目的是收起虚拟键盘。

在保持手机连接电脑的情况下，在 Xcode 中按快捷键“command + R”，即可将应用部署至手机，如图 8-10 所示。

图 8-10

8.3 Appium 的用法

8.3.1 准备

1. 环境搭建

关于环境搭建，本书不再赘述，有兴趣的读者可到博文官网下载查看。

2. 工程创建

打开 Mac 上安装的 Eclipse，创建一个新的 Maven 项目，关键信息填写如下。

Group Id：com.lujiatao

Artifact Id：iosapplicationtest

Name：iOS Application Test

当然，读者可根据实际情况填写，不需要和笔者填写的完全一致。

创建完成后，在 pom.xml 文件的<name>标签后输入以下粗体部分内容。

```
<project xmlns="http://maven.apache……"
    xmlns:xsi="http://www.w3……"
    xsi:schemaLocation="http://maven.apache…… http://maven.apache……">
    <modelVersion>4.0.0</modelVersion>

    <groupId>com.lujiatao</groupId>
    <artifactId>iosapplicationtest</artifactId>
    <version>0.0.1-SNAPSHOT</version>
    <name>iOS Application Test</name>

    <dependencies>
        <dependency>
            <groupId>io.appium</groupId>
            <artifactId>java-client</artifactId>
            <version>7.1.0</version>
        </dependency>
    </dependencies>

</project>
```

保存 pom.xml 文件，这时 Maven 会自动下载 Appium 客户端及其依赖的其他 jar 包。

8.3.2 初始化参数

初始化参数见表 7-1，这里不再赘述。

接下来看一个实例。在 src/test/java 中创建名为 com.lujiatao.iosapplicationtest 的 Package 及名为 IOSTest 的 Class（需要勾选“public static void main(String[] args)”），在 IOSTest 中输入以下代码：

```
package com.lujiatao.iosapplicationtest;

import java.net.MalformedURLException;
import java.net.URL;

import org.openqa.selenium.remote.DesiredCapabilities;

import io.appium.java_client.ios.IOSDriver;
import io.appium.java_client.ios.IOSElement;

public class IOSTest {

public static void main(String[] args) {
IOSDriver<IOSElement> driver = null;
DesiredCapabilities capabilities = new DesiredCapabilities();
capabilities.setCapability("platformName", "iOS");
capabilities.setCapability("platformVersion", "12.4");
capabilities.setCapability("deviceName", "iPhone 6s");
capabilities.setCapability("automationName", "XCUITest");
capabilities.setCapability("udid", "819fb7512554fc3248170a7093db7809cd1afd7f");
```

```
capabilities.setCapability("bundleId", "com.lujiatao.CalculatorForPPI");
try {
driver = new IOSDriver<IOSElement>(new URL("http://localhost:4723/wd/hub"),
        capabilities);
} catch (MalformedURLException e) {
e.printStackTrace();
}
}

}
```

保存代码，按组合键“fn + command + F11”运行工程，此时连接电脑的手机会自动启动待测应用并进入应用首页，如图 8-10 所示。需要说明的是，运行上述代码之前需保证手机屏幕已解锁，且 Appium 的服务器已启动。

下面对上述运行结果进行说明。

① 将启动参数的 platformName 设置为 iOS。platformVersion 和 deviceName 可以通过查看手机“设置 → 通用 → 关于本机”获得。automationName 填写 XCUITest，代表使用 XCUITest 进行自动化测试。udid 通过在终端执行 idevice_id -l 命令获得。bundleId 读者需根据实际情况替换。

② 构造一个 iOS 驱动，参数有两个。一个是 URL 对象，该对象描述了 Appium 服务器的地址，通过与服务器建立 HTTP 会话来传输数据；另一个是初始化参数对象 DesiredCapabilities。从这里可以看出，iOS 的初始化参数与 Android 的有一定差异。

8.3.3　元素操作

1. 元素查看

iOS 的元素查看同样不能直接在手机上进行，需要借助 Xcode 或者 Appium（推荐）。

（1）Xcode 查看

保持手机连接电脑，按快捷键“command + R”部署待测应用到手机上。单击 Xcode 左上角的调试导航及切换进程图标，如图 8-11 所示。

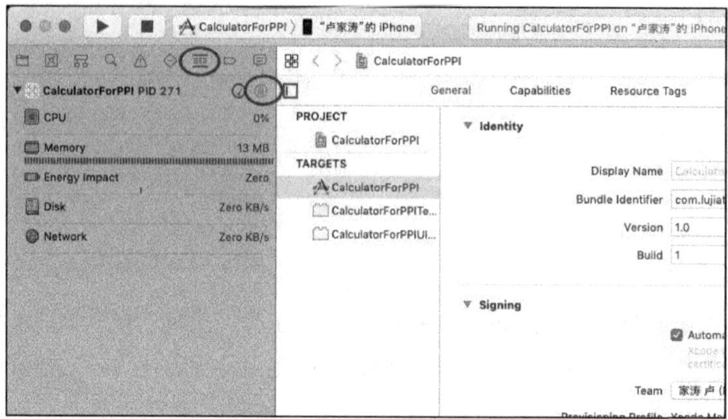

图 8-11

在打开的菜单中选择“View UI Hierarchy” 选项，元素布局便显示在了 Xcode 中，如图 8-12 所示。

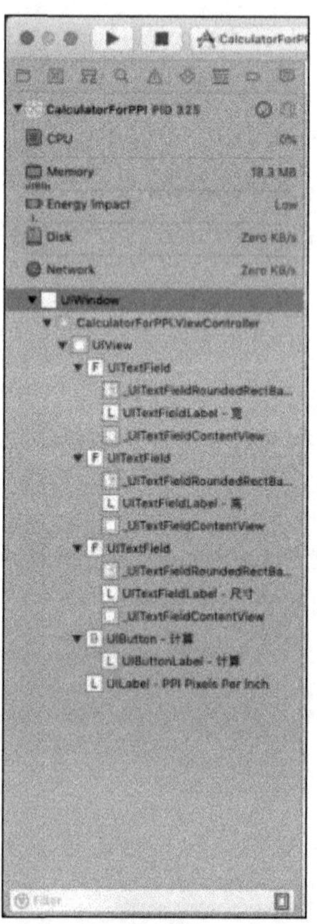

图 8-12

（2）Appium 查看

保持手机连接电脑，打开 Appium 服务器，单击右上角的放大镜图标，如图 8-13 所示。

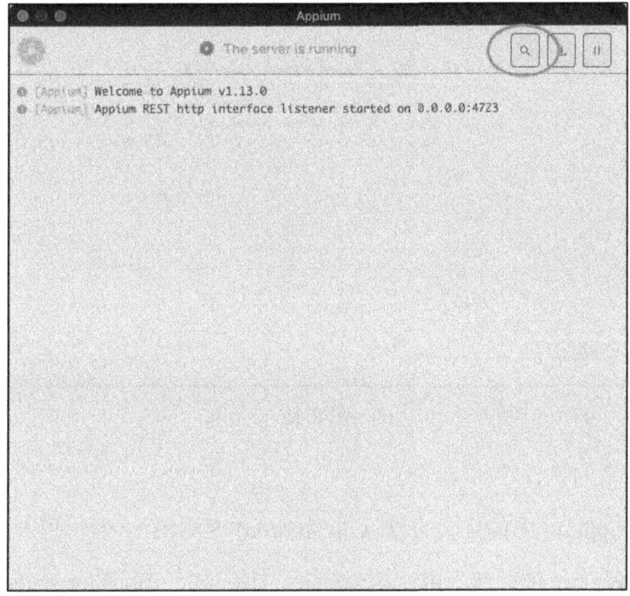

图 8-13

在 Automatic Server 中填写对应的 Desired Capabilities，如图 8-14 所示。

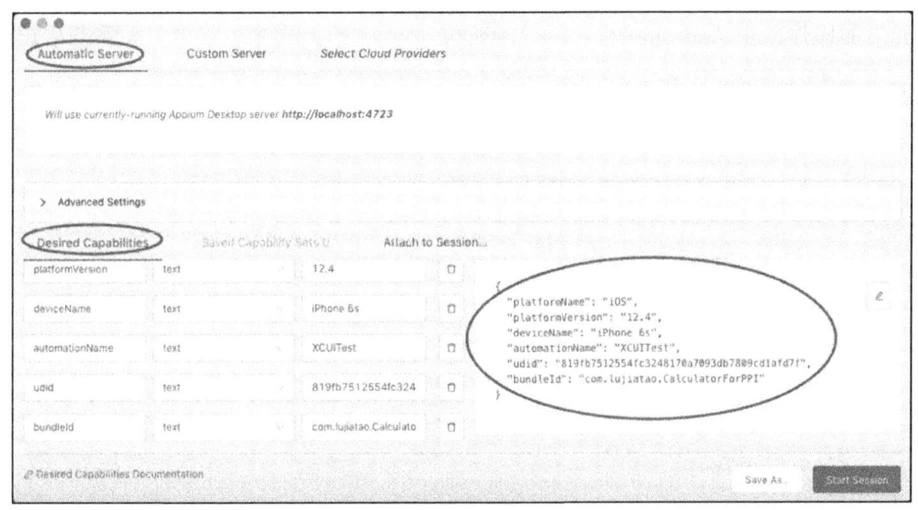

图 8-14

单击“Start Session”按钮，即可查看元素，如图 8-15 所示。

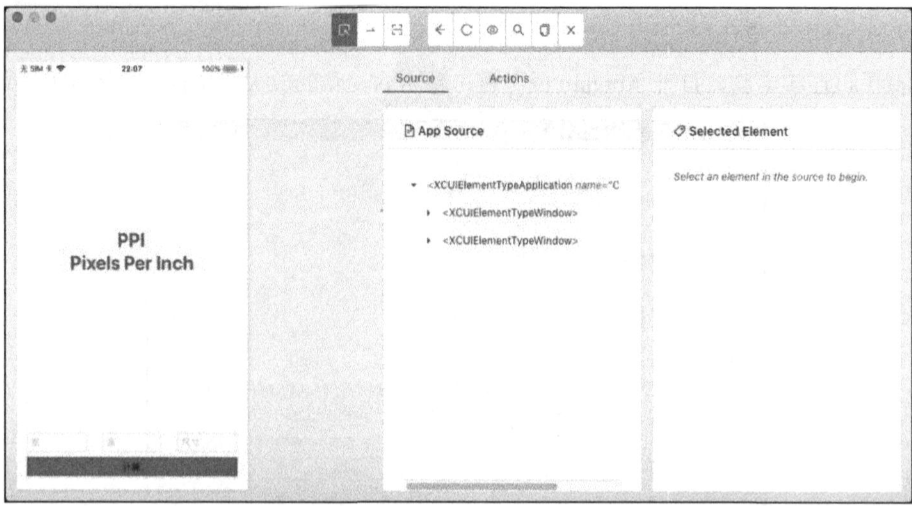

图 8-15

2. 定位元素

7.3.3 节介绍的 Appium 中的定位方法（非 Android 专用的）同样适用于 iOS，不再赘述。

iOS 中也有一些专用方法，比如 findElementByIosClassChain() 和 findElementByIosNsPredicate 等。

3. 操作元素

对元素的操作与在 Android 中类似，常见的有单击、输入文本、获取文本和获取属性等。

下面举例说明定位元素和操作元素。

在 IOSTest 中输入以下粗体部分代码：

```
package com.lujiatao.iosapplicationtest;

import java.net.MalformedURLException;
import java.net.URL;

import org.openqa.selenium.remote.DesiredCapabilities;

import io.appium.java_client.ios.IOSDriver;
import io.appium.java_client.ios.IOSElement;

public class IOSTest {

    public static void main(String[] args) {
        IOSDriver<IOSElement> driver = null;
        DesiredCapabilities capabilities = new DesiredCapabilities();
        capabilities.setCapability("platformName", "iOS");
```

```
        capabilities.setCapability("platformVersion", "12.4");
        capabilities.setCapability("deviceName", "iPhone 6s");
        capabilities.setCapability("automationName", "XCUITest");
        capabilities.setCapability("udid",
            "819fb7512554fc3248170a7093db7809cd1afd7f");
        capabilities.setCapability("bundleId", "com.lujiatao.CalculatorForPPI");
        try {
            driver = new IOSDriver<IOSElement>(new
                URL("http://localhost:4723/wd/hub"), capabilities);
        } catch (MalformedURLException e) {
            e.printStackTrace();
        }

        driver.findElementByXPath(
            "//XCUIElementTypeApplication[@name=\"CalculatorForPPI\"]/
            XCUIElementTypeWindow[1]/XCUIElementTypeOther/
            XCUIElementTypeTextField[1]")
                .sendKeys("750");
        driver.findElementByXPath(
            "//XCUIElementTypeApplication[@name=\"CalculatorForPPI\"]/
            XCUIElementTypeWindow[1]/XCUIElementTypeOther/
            XCUIElementTypeTextField[2]")
                .sendKeys("1334");
        driver.findElementByXPath(
            "//XCUIElementTypeApplication[@name=\"CalculatorForPPI\"]/
            XCUIElementTypeWindow[1]/XCUIElementTypeOther/
            XCUIElementTypeTextField[3]")
                .sendKeys("4.7");
        driver.findElementByName("Done").click();
        driver.findElementByName("计算").click();
        String result = driver.findElementByXPath("//XCUIElementTypeStaticText").
            getText();
        System.out.println(result);
    }

}
```

保存代码，按组合键“fn + command + F11”运行工程， Eclipse 的控制台输出如下：

```
九月 05, 2019 11:00:38 下午 io.appium.java_client.remote.AppiumCommandExecutor$1
lambda$0
信息: Detected dialect: W3C
326 ppi
```

此时待测应用会自动打开，并显示如图 8-16 所示的内容。

图 8-16

下面对上述运行结果进行说明。

① 在 Appium 中单击左边的输入框，右边即可显示元素的信息，可以看到输入框的 XPath 值，如图 8-17 所示。

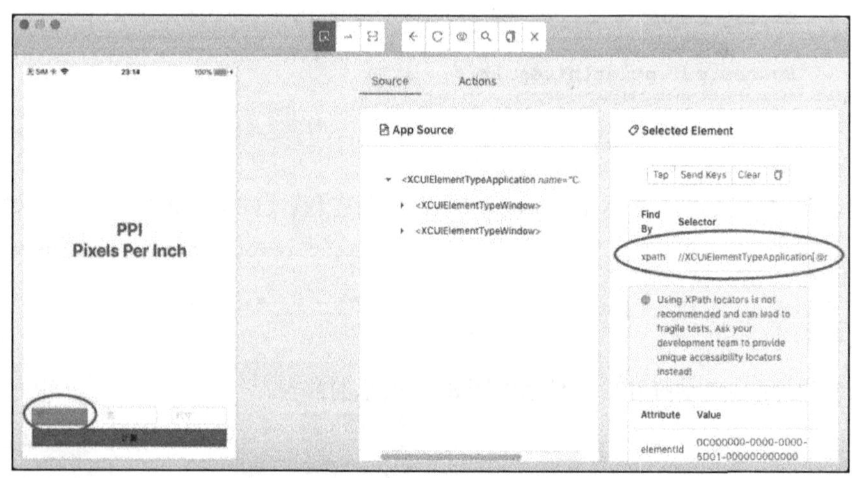

图 8-17

用同样的方式可以找到 UILabel 的 XPath 路径以及“计算”按钮的 Name 值。

② 这里单击了 Name 值为“Done”的元素，这个元素是虚拟键盘上的“Done”按钮。如果输入后不收起虚拟键盘，那么虚拟键盘会将“计算”按钮挡住，如图 8-18 所示。

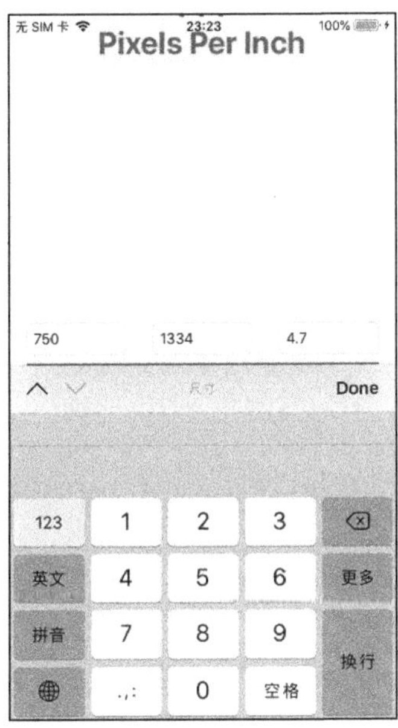

图 8-18

③ 使用 getText()方法获取元素文本，并将文本打印到控制台，文本对应 Appium 中显示的 Name 值。

8.3.4 应用操作

在 iOS 中，常见的应用操作与 Android 类似，参见 7.3.4 节。下面仍以打开待测应用并继续打开手机自带的计算器为例。

删除 IOSTest 中的内容，输入以下代码：

```
package com.lujiatao.iosapplicationtest;

import java.net.MalformedURLException;
import java.net.URL;

import org.openqa.selenium.remote.DesiredCapabilities;

import io.appium.java_client.ios.IOSDriver;
import io.appium.java_client.ios.IOSElement;

public class IOSTest {

    public static void main(String[] args) {
```

```
        IOSDriver<IOSElement> driver = null;
        DesiredCapabilities capabilities = new DesiredCapabilities();
        capabilities.setCapability("platformName", "iOS");
        capabilities.setCapability("platformVersion", "12.4");
        capabilities.setCapability("deviceName", "iPhone 6s");
        capabilities.setCapability("automationName", "XCUITest");
        capabilities.setCapability("udid",
           "819fb7512554fc3248170a7093db7809cd1afd7f");
        capabilities.setCapability("bundleId", "com.lujiatao.CalculatorForPPI");
        try {
            driver = new IOSDriver<IOSElement>(new
                 URL("http://localhost:4723/wd/hub"), capabilities);
        } catch (MalformedURLException e) {
            e.printStackTrace();
        }
        boolean isInstalled = driver.isAppInstalled("com.apple.calculator");
        if (isInstalled) {
            driver.activateApp("com.apple.calculator");
        }
    }

}
```

保存代码，按组合键“fn + command + F11”运行工程，此时待测应用会自动打开，同时打开了手机自带的计算器应用，如图 8-19 所示。

图 8-19

这里先判断是否安装了计算器应用，如果已安装，则启动应用。可执行下面的命令查看手机中安装的所有应用的 Bundle ID，执行结果如图 8-20 所示。

```
ios-deploy --id [udid] --list_bundle_id
```

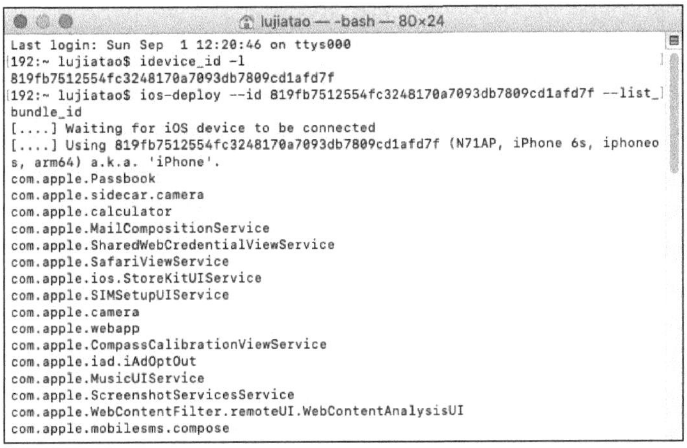

```
lujiatao — -bash — 80×24
Last login: Sun Sep  1 12:20:46 on ttys000
192:~ lujiatao$ idevice_id -l
819fb7512554fc3248170a7093db7809cd1afd7f
192:~ lujiatao$ ios-deploy --id 819fb7512554fc3248170a7093db7809cd1afd7f --list_
bundle_id
[....] Waiting for iOS device to be connected
[....] Using 819fb7512554fc3248170a7093db7809cd1afd7f (N71AP, iPhone 6s, iphoneo
s, arm64) a.k.a. 'iPhone'.
com.apple.Passbook
com.apple.sidecar.camera
com.apple.calculator
com.apple.MailCompositionService
com.apple.SharedWebCredentialViewService
com.apple.SafariViewService
com.apple.ios.StoreKitUIService
com.apple.SIMSetupUIService
com.apple.camera
com.apple.webapp
com.apple.CompassCalibrationViewService
com.apple.iad.iAdOptOut
com.apple.MusicUIService
com.apple.ScreenshotServicesService
com.apple.WebContentFilter.remoteUI.WebContentAnalysisUI
com.apple.mobilesms.compose
```

图 8-20

8.3.5　系统操作

在 iOS 中，常见的系统操作与 Android 类似，参见 7.3.5 节。下面仍以锁屏为例，将 IOSTest 的 main()方法中最后 4 行代码替换为以下代码：

```
boolean isLocked = driver.isDeviceLocked();
if (!isLocked) {
    driver.lockDevice();
}
```

保存代码，按组合键“fn + command + F11”运行工程，此时待测应用会自动打开，打开后自动锁屏。

8.3.6　使用 iOS 模拟器

之前的 iOS 应用测试都是基于手机进行的，在 iOS 应用开发过程中，经常会使用模拟器进行调试，当然，我们的 iOS 应用测试也可以使用模拟器进行测试。

在 Xcode 左上角单击手机即可切换成模拟器，如图 8-21 所示。

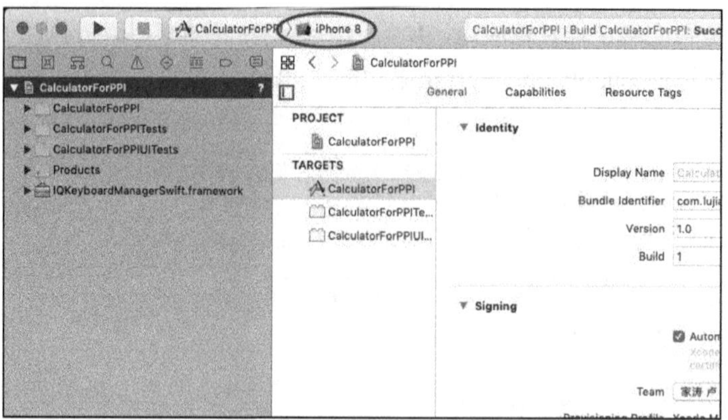

图 8-21

切换后按快捷键“command + R”将待测应用部署到模拟器上， Xcode 会自动启动模拟器并打开待测应用。

在终端中执行以下命令将 WebDriverAgentRunner 部署到模拟器上。

```
cd
/Applications/Appium.app/Contents/Resources/app/node_modules/appium/node_module
s/appium-xcuitest-driver/WebDriverAgent
xcodebuild -project WebDriverAgent.xcodeproj -scheme WebDriverAgentRunner
-destination 'id=258E2539-F643-48C8-AA7D-4E762C9DF720' test
```

模拟器的 deviceName 为“Simulator”。可执行以下命令获取 udid，执行后如图 8-22 所示。

```
instruments -s devices
```

```
lujiatao — -bash — 80×24
iPad Pro (12.9-inch) (2nd generation) (12.4) [11055414-711A-41F4-B247-66BFD74973
DD] (Simulator)
iPad Pro (12.9-inch) (3rd generation) (12.4) [4DA8679A-4B6F-44F6-A760-4BFB7A209E
46] (Simulator)
iPad Pro (9.7-inch) (12.4) [BA8F40BD-79D5-46A1-92CA-090FDD0991CF] (Simulator)
iPhone 5s (12.4) [516C6341-10A8-4092-97D3-73DCC15A6B53] (Simulator)
iPhone 6 (12.4) [8EF831BB-A5BC-44C3-A2E4-E6C8D6AE88AC] (Simulator)
iPhone 6 Plus (12.4) [9B165FCF-EBB5-49D6-9BC1-3ABF84F84BAD] (Simulator)
iPhone 6s (12.4) [7AFF6C37-67C7-4272-98EA-4B0199D84E9E] (Simulator)
iPhone 6s Plus (12.4) [03B544B5-905A-4931-A93D-BC1BC010CF8D] (Simulator)
iPhone 7 (12.4) [3DF46042-60AC-4B90-8C69-D23C049082DB] (Simulator)
iPhone 7 Plus (12.4) [18E759AA-A803-447E-AC37-166EFD20B22F] (Simulator)
iPhone 8 (12.4) [258E2539-F643-48C8-AA7D-4E762C9DF720] (Simulator)
iPhone 8 Plus (12.4) [9BEF41BE-5E2D-4AF6-ADC1-C01FC75CAAFB] (Simulator)
iPhone SE (12.4) [43D0EA40-1535-4CB5-B327-F6ED81F50290] (Simulator)
iPhone X (12.4) [7544C781-15B5-4BAB-8673-86F1D006716A] (Simulator)
iPhone Xs (12.4) [7BE63361-9B66-4971-9B76-ADCEDA7A2E4E] (Simulator)
iPhone Xs (12.4) + Apple Watch Series 4 - 40mm (5.3) [504733E7-9CA7-40D8-887A-9F
DC8BAF711C] (Simulator)
iPhone Xs Max (12.4) [012078B5-02AF-4BE3-A92D-E20295603420] (Simulator)
iPhone Xs Max (12.4) + Apple Watch Series 4 - 44mm (5.3) [3DE91199-02D4-4216-8FE
2-145A6213604C] (Simulator)
iPhone XR (12.4) [24351912-F21F-4902-B701-2515D3847F8E] (Simulator)
192:~ lujiatao$
```

图 8-22

将测试代码中的手机 udid 和 deviceName 替换成模拟器的 udid 和 deviceName，便可以使用模拟器进行自动化测试了。

8.4 TestNG 集成 Appium

本节使用 TestNG 集成 Appium 演示一个完整的 iOS 自动化测试用例，该用例的功能为启动待测应用，计算 iPhone 6S 的 PPI，并与预期值做对比。

在 pom.xml 文件的<dependencies>标签中输入以下粗体部分内容。

```
<project xmlns="http://maven.apache……"
    xmlns:xsi="http://www.w3……"
    xsi:schemaLocation="http://maven.apache…… http://maven.apache……">
    <modelVersion>4.0.0</modelVersion>

    <groupId>com.lujiatao</groupId>
    <artifactId>iosapplicationtest</artifactId>
    <version>0.0.1-SNAPSHOT</version>
    <name>iOS Application Test</name>

    <dependencies>
        <dependency>
            <groupId>io.appium</groupId>
            <artifactId>java-client</artifactId>
            <version>7.1.0</version>
        </dependency>
        <dependency>
            <groupId>org.testng</groupId>
            <artifactId>testng</artifactId>
            <version>6.14.3</version>
            <scope>test</scope>
        </dependency>
    </dependencies>

</project>
```

保存 pom.xml 文件，这时 Maven 会自动下载 TestNG 及其依赖的其他 jar 包。

依赖 jar 包下载完成后，参照 2.2.3 节安装 TestNG 插件。安装后在工程（iosapplicationtest）上用鼠标右击，从弹出的快捷菜单中选择“TestNG → Convert to TestNG” 选项，在工程中生成 testng.xml 文件。需要注意的是，testng.xml 文件的<!DOCTYPE>标签中默认协议为 HTTPS，需要改成 HTTP，否则在运行 TestNG 时会报错“unable to find valid certification path to requested target”。

删除 IOSTest 中的内容，输入以下代码：

```
package com.lujiatao.iosapplicationtest;

import org.testng.annotations.AfterClass;
import org.testng.annotations.Test;
import org.testng.annotations.BeforeClass;
```

```
import org.testng.AssertJUnit;
import java.net.MalformedURLException;
import java.net.URL;

import org.openqa.selenium.remote.DesiredCapabilities;

import io.appium.java_client.ios.IOSDriver;
import io.appium.java_client.ios.IOSElement;

public class IOSTest {

    private IOSDriver<IOSElement> driver = null;

    @BeforeClass
    public void init() {
        DesiredCapabilities capabilities = new DesiredCapabilities();
        capabilities.setCapability("platformName", "iOS");
        capabilities.setCapability("platformVersion", "12.4");
        capabilities.setCapability("deviceName", "iPhone 6s");
        capabilities.setCapability("automationName", "XCUITest");
        capabilities.setCapability("udid",
            "819fb7512554fc3248170a7093db7809cd1afd7f");
        capabilities.setCapability("bundleId", "com.lujiatao.CalculatorForPPI");
        try {
            driver = new IOSDriver<IOSElement>(new
                URL("http://localhost:4723/wd/hub"), capabilities);
        } catch (MalformedURLException e) {
            e.printStackTrace();
        }
    }

    @Test
    public void testCase1() {
        driver.findElementByXPath(
            "//XCUIElementTypeApplication[@name=\"CalculatorForPPI\"]/
                XCUIElementTypeWindow[1]/XCUIElementTypeOther/XCUIElementTypeTe
                xtField[1]").sendKeys("750");
        driver.findElementByXPath(
            "//XCUIElementTypeApplication[@name=\"CalculatorForPPI\"]/
                XCUIElementTypeWindow[1]/XCUIElementTypeOther/XCUIElementTypeTe
                xtField[2]").sendKeys("1334");
        driver.findElementByXPath(
            "//XCUIElementTypeApplication[@name=\"CalculatorForPPI\"]/
                XCUIElementTypeWindow[1]/XCUIElementTypeOther/XCUIElementTypeTe
                xtField[3]").sendKeys("4.7");
        driver.findElementByName("Done").click();
        driver.findElementByName("计算").click();
        String expected = "326 ppi";
        String actual = driver.findElementByXPath("//XCUIElementTypeStaticText").
            getText();
        AssertJUnit.assertEquals(expected, actual);
```

```
    }

    @AfterClass
    public void clear() {
        driver.quit();
    }

}
```

修改 testng.xml 文件，在<test>标签中新增以下粗体部分内容：

```
<?xml version="1.0" encoding="UTF-8"?>
<!DOCTYPE suite SYSTEM "http://testng.org/testng-1.0.dtd">
<suite name="Suite">
    <test thread-count="5" name="Test">
        <classes>
            <class name="com.lujiatao.iosapplicationtest.IOSTest" />
        </classes>
    </test> <!-- Test -->
</suite> <!-- Suite -->
```

保存所做的修改，在 testng.xml 上用鼠标右击，从弹出的快捷菜单中选择"Run As → TestNG Suite" 选项，查看测试报告，如图 8-23 所示。

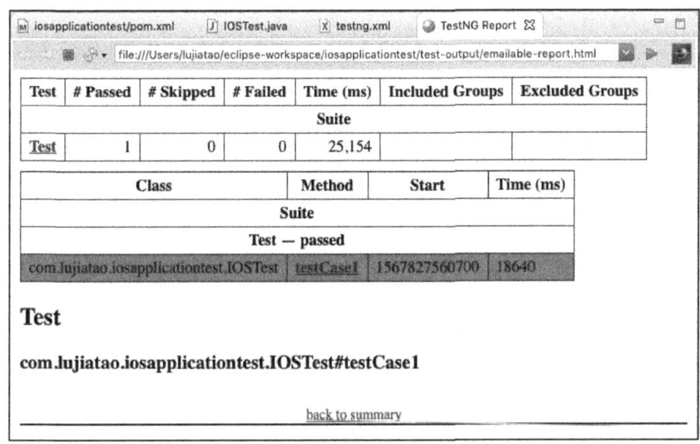

Test	# Passed	# Skipped	# Failed	Time (ms)	Included Groups	Excluded Groups
Suite						
Test	1	0	0	25,154		

Class	Method	Start	Time (ms)
Suite			
Test — passed			
com.lujiatao.iosapplicationtest.IOSTest	testCase1	1567827560700	18640

Test

com.lujiatao.iosapplicationtest.IOSTest#testCase1

back to summary

图 8-23

现对以上运行结果进行说明。

① init()方法初始化了一个 iOS 驱动对象，并在测试用例方法（testCase1）中使用，最后在 clear()方法中关闭。

② 通过 XPath 和 Name 等方法定位和操作元素。

③ 使用 TestNG 的断言方法对预期和实际结果进行比较完成断言。

第9章 自动化测试实战

通过前面的介绍，相信读者已经对单元、接口和界面自动化测试技术有了深刻的认识，本章通过两个实战项目对前面的知识做进一步的巩固。

9.1 实战项目部署安装

9.1.1 JForum 论坛部署

JForum 为著名的开源论坛项目，其支持多种语言，并使用了 BSD 开源协议。

1．安装 JDK

2．安装 Tomcat

访问官方下载页面下载 apache-tomcat-9.0.27.exe 文件。

双击 apache-tomcat-9.0.27.exe，按照提示安装即可。安装路径可以自行修改，笔者将 Tomcat 安装在了 D:\Program Files 目录。安装完成后访问 http://localhost:8080/，如图 9-1 所示则表示安装成功。

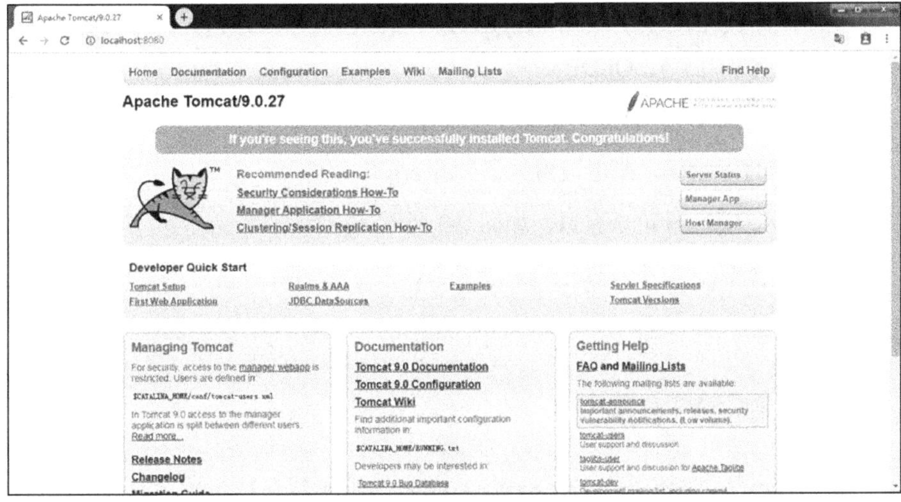

图 9-1

3. 安装 JForum

从官方下载 jforum-2.6.2.war 文件。

将 jforum-2.6.2.war 文件重命名为 jforum.war，并拷贝到 Tomcat 的 webapps 目录中，重启 Tomcat。重启方法是双击任务栏的 Tomcat 图标，打开如图 9-2 所示对话框，依次单击“Stop → Start”即可。

图 9-2

访问 http://localhost:8080/jforum/install.jsp，进入部署页面。按照提示完成 JForum 论坛部署即可。有以下几点需要注意。

① 出于学习目的数据库最好采用 HSQLDB，否则需要单独搭建数据库服务，增加环境搭建成本，笔者这里就采用了 HSQLDB。

② 数据库编码选择 UTF-8。

③ 设置系统管理员密码，笔者设置为了“qwerty”。

安装完成后进入 JForum 论坛首页，如图 9-3 所示。

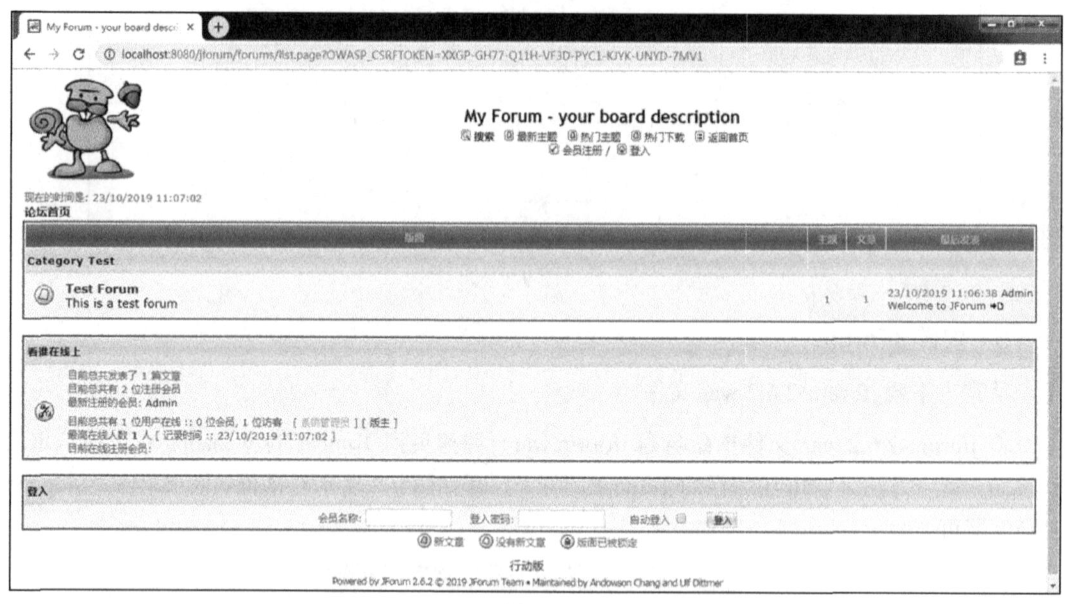

图 9-3

4．初始化操作

为了更好地满足实战需求，我们需要对 JForum 论坛进行一些必要的初始化操作。使用管理员账号和密码登录，并单击页面底部的“进入后台管理”进入管理，进行以下操作。

① 进入“系统设置”，修改论坛名称为“手机论坛”，修改论坛页面标题为“手机论坛”。

② 进入“版面分类”，修改默认分类 Category Test 为“Apple”，并新增“HUAWEI”分类。新增时在“选择允许存取此新增分类的会员群组”选项中勾选 Administration 和 General 选项。

③ 进入“版面管理”，修改默认版面 Test Forum 为“iPhone 11”，并在“Apple”分类下新增“iPhone 11 Pro”和“iPhone 11 Pro Max”版面，在“HUAWEI”分类下新增“Mate 30”和“Mate 30 Pro”版面。新增时“版面说明”可以随意填写，比如在“iPhone 11”版面，笔者填写的是“iPhone 11 讨论区”，其他版面以此类推。

④ 进入“会员分组”，设置“General”权限，在“匿名发帖”选项中选中所有版面，在“限制回复”和“附件相关”选项中选中“允许全部”。

回到论坛主页，论坛已经更新为初始化后的样子，如图 9-4 所示。

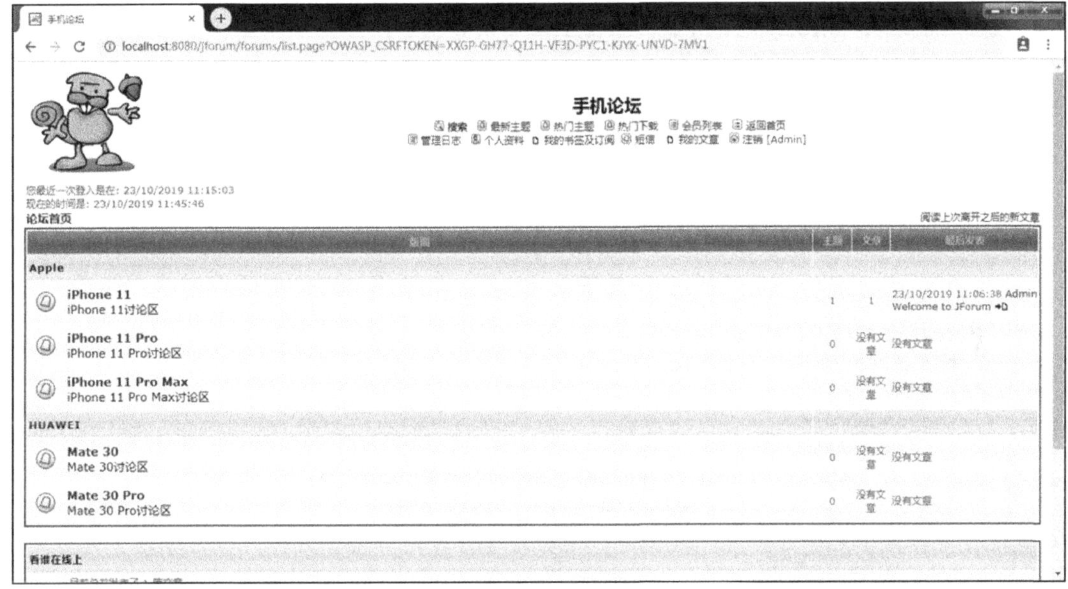

图 9-4

9.1.2　AnExplorer 文件管理器安装

AnExplorer 是 Android 平台的一个多合一文件管理器，支持基于 Android 的手机、平板电脑、智能手表和智能电视等。AnExplorer 使用了 Android Material 设计风格，目前在 GitHub 上拥有 1000 多个 Star，使用的开源协议为 Apache 2.0。

由于国内无法直接访问 Google Paly，因此需要在 AnExplorer 的 GitHub 页面下载安装文件。打开 GitHub 主页，搜索 AnExplorer 即可。

9.2　Web 自动化测试实战

9.2.1　分层和解耦

1. 工程创建

创建一个新的 Maven 项目，关键信息填写如下。

Group Id：com.lujiatao

Artifact Id：webtestinaction

Name：Web Test In Action

读者可根据实际情况填写，不需要和笔者填写的完全一致。

创建完成后，在 pom.xml 文件的<name>标签后输入以下粗体部分内容：

```
<project xmlns="http://maven.apache……"
    xmlns:xsi="http://www.w3……"
    xsi:schemaLocation="http://maven.apache…… http://maven.apache……">
    <modelVersion>4.0.0</modelVersion>

    <groupId>com.lujiatao</groupId>
    <artifactId>webtestinaction</artifactId>
    <version>0.0.1-SNAPSHOT</version>
    <name>Web Test In Action</name>

    <dependencies>
        <dependency>
            <groupId>org.seleniumhq.selenium</groupId>
            <artifactId>selenium-java</artifactId>
            <version>3.141.59</version>
        </dependency>
        <dependency>
            <groupId>org.apache.poi</groupId>
            <artifactId>poi-ooxml</artifactId>
            <version>4.1.1</version>
        </dependency>
        <dependency>
            <groupId>org.testng</groupId>
            <artifactId>testng</artifactId>
            <version>6.14.3</version>
            <scope>test</scope>
        </dependency>
    </dependencies>

</project>
```

保存 pom.xml 文件，这时 Maven 会自动下载 Selenium、poi-ooxml、TestNG 及其依赖的其他 jar 包。其中，poi-ooxml 为 Excel 文件处理工具，后续会用到。

当依赖 jar 包下载完成后，在工程（webtestinaction）上单击右键，选择 TestNG → Convert to TestNG 在工程中生成的 testng.xml 文件。

2．构建工程结构

按照如图 9-5 所示新建对应 Package 或目录。

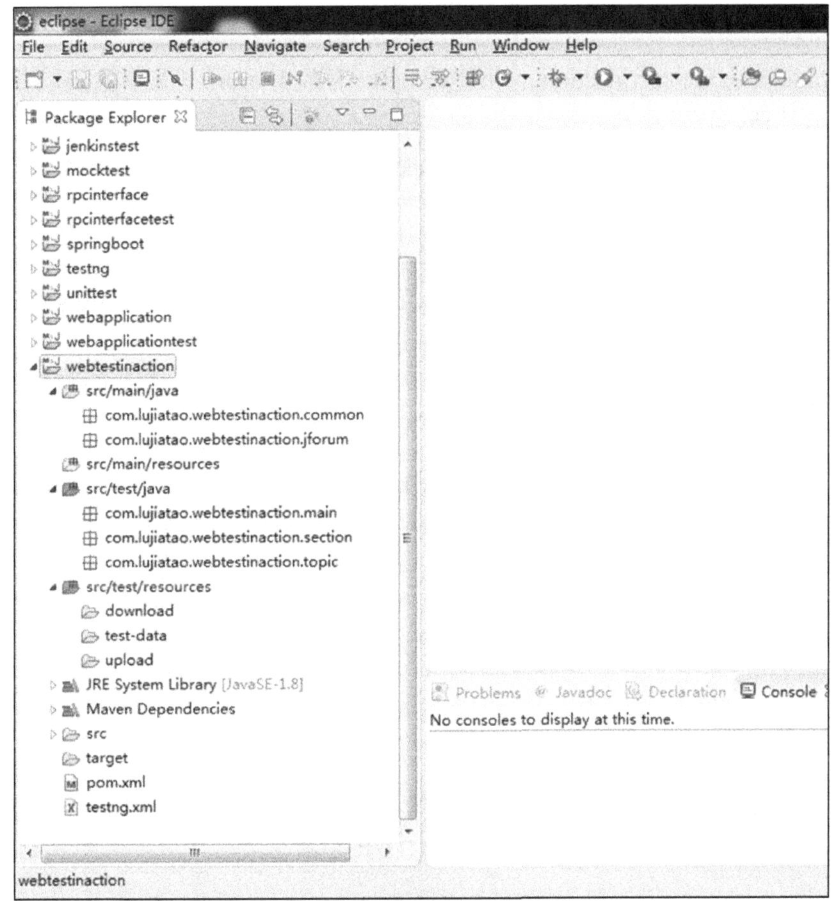

图 9-5

该工程结构采用了分层和解耦策略，详细解释如下。

（1）src/main/java

- com.lujiatao.webtestinaction.common Package：存放公共函数封装，该部分代码与具体业务无关，适用于所有项目，可单独打包在所有项目中引用。
- com.lujiatao.webtestinaction.jforum Package：存放业务函数封装，在本节中代表 JForum 论坛，该部分代码与具体业务有关，仅适用于当前项目。读者可能会有疑问，既然仅适用于 JForum 论坛为何不直接放在 src/test/java 中？假设某公司有多个项目，且项目之间可能有交互操作，那么业务函数实际上是可以复用的，这种情况下也可单独打包在所有项目中，以便被复用。

在分层策略中，上述 Package 中的函数代表逻辑层。

（2）src/test/java

- com.lujiatao.webtestinaction.main Package：存放 JForum 论坛首页自动化测试用例，包含注册、登录和搜索等功能。
- com.lujiatao.webtestinaction.section Package：存放 JForum 论坛版面自动化测试用例，包括搜索、发帖和切换版面等功能。
- com.lujiatao.webtestinaction.topic Package：存放 JForum 论坛帖子页面自动化测试用例，包括回帖、引用、编辑和删除等功能。

以上 Package 中存放自动化测试用例，理论上自动化测试用例中只能调用 TestNG、公共函数封装和业务函数封装中的 API 以达到严格的“分层”，但实际项目中自动化测试用例往往会有其他代码。

在分层策略中以上 Package 中的自动化测试用例代表用例层。

（3）src/test/resources

download 目录：存放下载文件。

test-data 目录：存放测试数据文件，可采用.xlsx、.xls 或.csv 等文件存放测试数据。

upload 目录：存放上传文件。

在分层策略中，上述目录中的文件代表数据层。

至此，通过逻辑层、用例层和数据层的三层划分实现了对整个自动化测试工程的最基本的解耦，更细粒度的解耦会在自动化测试用例的设计中体现。

9.2.2 公共函数和业务函数封装

1. 公共函数封装

公共函数与具体业务无关，包括但不限于操作 Linux、数据库、文件和时间等的函数。

在本节的实战项目中笔者封装了一个 readExcel()方法，用于读取 Excel 文件的内容。在 com.lujiatao.webtestinaction.common Package 中创建名为 Files 的 Class，在 Files 中输入以下代码：

```
package com.lujiatao.webtestinaction.common;

import java.io.File;
import java.io.FileInputStream;
import java.util.ArrayList;
import java.util.HashMap;

import org.apache.poi.hssf.usermodel.HSSFSheet;
import org.apache.poi.hssf.usermodel.HSSFWorkbook;
import org.apache.poi.poifs.filesystem.POIFSFileSystem;
```

```
import org.apache.poi.ss.usermodel.Row;
import org.apache.poi.xssf.usermodel.XSSFSheet;
import org.apache.poi.xssf.usermodel.XSSFWorkbook;

public class Files {

    /***
     * 读取 Excel 文件
     *
     * @param dir       文件目录
     * @param file      文件名
     * @param sheetName Sheet 名
     * @return
     */
    public static ArrayList<HashMap<String, String>> readExcel(String dir, String
        file, String sheetName) {
        ArrayList<HashMap<String, String>> result = new ArrayList<>();
        if (file.endsWith(".xlsx")) {
            try {
                XSSFWorkbook book = new XSSFWorkbook(new File(dir + File.separator
                    + file));
                XSSFSheet sheet = book.getSheet(sheetName);
                HashMap<String, String> hm;
                Row titles = sheet.getRow(0);
                if (sheet.getLastRowNum() != 0) {// 排除只有表头的 Sheet
                    Row row;
                    for (int i = 1; i < sheet.getLastRowNum() + 1; i++) {
                        hm = new HashMap<String, String>();
                        row = sheet.getRow(i);
                        for (int j = 0; j < titles.getLastCellNum(); j++) {
                            hm.put(titles.getCell(j).getStringCellValue(),
                               row.getCell(j).getStringCellValue());
                        }
                        result.add(hm);
                    }
                } else {// 处理只有表头的 Sheet
                    hm = new HashMap<String, String>();
                    for (int i = 0; i < titles.getLastCellNum(); i++) {
                        hm.put(titles.getCell(i).getStringCellValue(), "");
                    }
                    result.add(hm);
                }
                book.close();
            } catch (Exception e) {
                e.printStackTrace();
            }
        } else {
            try {
                HSSFWorkbook book = new HSSFWorkbook(
```

```
                    new POIFSFileSystem(new FileInputStream(new File(dir +
                        File.separator + file))));
                HSSFSheet sheet = book.getSheet(sheetName);
                HashMap<String, String> hm;
                Row titles = sheet.getRow(0);
                if (sheet.getLastRowNum() != 0) {// 排除只有表头的 Sheet
                    Row row;
                    for (int i = 1; i < sheet.getLastRowNum() + 1; i++) {
                        hm = new HashMap<String, String>();
                        row = sheet.getRow(i);
                        for (int j = 0; j < titles.getLastCellNum(); j++) {
                            hm.put(titles.getCell(j).getStringCellValue(),
                                row.getCell(j).getStringCellValue());
                        }
                        result.add(hm);
                    }
                } else {// 处理只有表头的 Sheet
                    hm = new HashMap<String, String>();
                    for (int i = 0; i < titles.getLastCellNum(); i++) {
                        hm.put(titles.getCell(i).getStringCellValue(), "");
                    }
                    result.add(hm);
                }
                book.close();
            } catch (Exception e) {
                e.printStackTrace();
            }
        }
        return result;
    }

}
```

下面对上述代码进行说明。

① 使用 HashMap<String, String>存放 Excel 中每一行的内容（Key 为表头，Value 为值），将每一行的内容放进 ArrayList 中，这样可以保证内容的有序性。因此整个 readExcel()方法的返回值为 ArrayList<HashMap<String, String>>。

② readExcel()方法接收 3 个参数，第一个参数为文件路径；第二个参数为文件名，文件名包括文件的扩展名，即.xlsx 或.xls；第三个参数为 Sheet 名，该方法每次只能读取一个 Sheet 的内容，因此需要指定 Sheet 名。

③ 由于.xlsx 和.xls 文件的处理方法有区别，因此需要分别对待。.xlsx 文件使用 XSSFWorkbook 和 XSSFSheet 处理，而.xls 文件使用 HSSFWorkbook 和 HSSFSheet 处理。

④ 针对 Sheet 中有值（不只是表头）的情况，遍历每一行，将表头作为 Key，将遍历的行作为 Value 存到 HashMap<String, String>中，再放进 ArrayList。

⑤ 针对 Sheet 中无值（只有表头）的情况，将表头作为 Key，将空字符串作为 Value 存到 HashMap<String, String>中，再放进 ArrayList。

⑥ XSSFWorkbook 和 HSSFWorkbook 使用完成后均需要关闭。另外，整个过程可能会产生异常，需要在代码中对异常进行捕获和处理。

2. 业务函数封装

业务函数与具体业务有关，针对不同的项目业务，函数有很大的差别，因此也注定了业务函数的复用性远远不及公共函数。

在本节的实战项目中，笔者封装了 login()、goSection()、goAddTopic()和 addTopic()共 4 个方法。在 com.lujiatao.webtestinaction.jforum Package 中创建名为 JForum 的 Class，在 JForum 中输入以下代码：

```
package com.lujiatao.webtestinaction.jforum;

import java.util.List;

import org.openqa.selenium.By;
import org.openqa.selenium.WebDriver;
import org.openqa.selenium.WebElement;

public class JForum {

    /***
     * 登录
     *
     * @param driver    浏览器驱动
     * @param url       URL
     * @param username  用户名
     * @param password  密码
     * @param autoLogin 是否自动登录
     * @return
     */
    public static boolean login(WebDriver driver, String url, String username,
        String password, boolean autoLogin) {
        driver.get(url);
        driver.findElement(By.id("login")).click();
        driver.findElement(By.name("username")).sendKeys(username);
        driver.findElement(By.name("password")).sendKeys(password);
        if (autoLogin) {
            driver.findElement(By.id("autologin")).click();
        }
        driver.findElement(By.name("login")).click();
        return driver.findElement(By.id("logout")).getText().equals("注销 [" +
            username + "]");
```

```
}

/***
 * 进入版面
 *
 * @param driver  浏览器驱动
 * @param section 版面名称
 * @return
 */
public static boolean goSection(WebDriver driver, String section) {
    List<WebElement> sections =
        driver.findElements(By.cssSelector("a.forumlink"));
    for (WebElement tmp : sections) {
        if (tmp.getText().equals(section)) {
            tmp.click();
            return driver.findElements(By.className("maintitle")).
                get(1).getText().equals(section);
        }
    }
    return false;
}

/***
 * 进入发帖页面
 *
 * @param driver 浏览器驱动
 * @return
 */
public static boolean goAddTopic(WebDriver driver) {
    driver.findElements(By.className("icon_new_topic")).get(0).click();
    return driver.findElement(By.xpath("//th[@class='thhead']/b")).
        getText().trim().equals("发表新主题");
}

/***
 * 发帖
 *
 * @param driver  浏览器驱动
 * @param title   帖子标题
 * @param content 帖子内容
 * @return
 */
public static boolean addTopic(WebDriver driver, String title, String content) {
    driver.findElement(By.name("subject")).sendKeys(title);
    driver.findElement(By.name("message")).sendKeys(content);
    driver.findElement(By.id("btnSubmit")).click();
    if (driver.findElement(By.xpath("//div[@class='subject']/a")).
        getText().equals(title)
            && driver.findElement(By.xpath("//div[@class='postbody']/div")).
```

```
                    getText().equals(content)) {
                return true;
            }
            return false;
        }

    }
```

下面对上述代码进行说明。

login()方法为登录方法，通过传入浏览器驱动、URL、用户名、密码和是否自动登录 5 个参数进行登录操作，最后匹配指定字符串。若匹配成功，则返回 true，否则返回 false。是否自动登录是在 JForum 论坛登录页，由读者决定是否勾选，如图 9-6 所示。

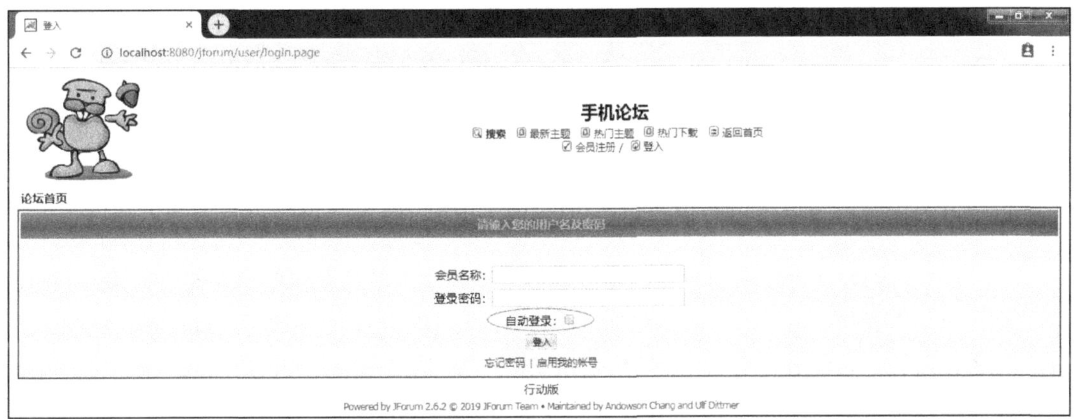

图 9-6

① goSection()方法为进入指定版面的方法，传入浏览器驱动及想要进入的版面名称即可。在 goSection()方法中，首先使用 CSS 定位选择所有满足条件的版面，然后遍历所有版面，找到想要进入的版面。

② goAddTopic()方法为进入发帖页面的方法。实际上发帖页面有两个“发表主题”按钮，如图 9-7 所示。在 goAddTopic()方法中单击的是第一个，即 index 为 0 的那个。

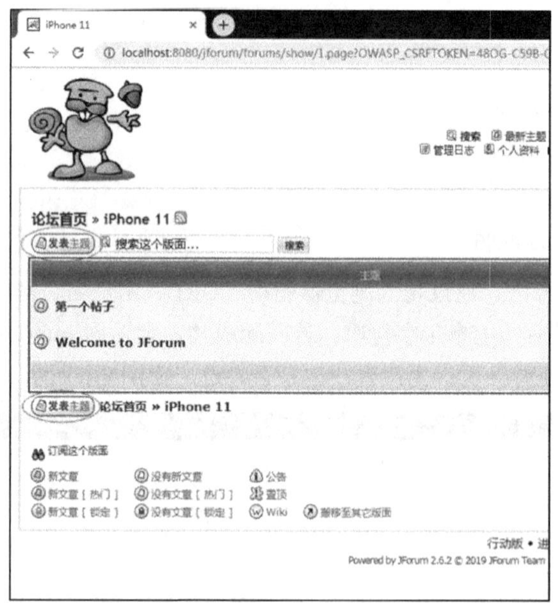

图 9-7

③ addTopic()方法为发帖的方法，除浏览器驱动外，入参只有标题和内容。实际上，在发帖页面还有很多操作，这里进行了简化处理。与前面几个方法不一样的是，addTopic()方法最后的返回值是对两个条件进行判断（发帖标题和发帖内容），当两个条件同时满足时才返回 true。

9.2.3　自动化测试用例编写

在本节的实战项目中，笔者编写了两个自动化测试场景，即登录和发帖。其中，发帖是依赖登录的，因此在发帖的自动化测试用例中也会执行登录操作。

首先在 com.lujiatao.webtestinaction.main Package 中创建名为 Account 的 Class，在 Account 中输入以下代码：

```
package com.lujiatao.webtestinaction.main;

import java.io.File;
import java.util.ArrayList;
import java.util.HashMap;

import org.openqa.selenium.WebDriver;
import org.openqa.selenium.chrome.ChromeDriver;
import org.testng.Assert;
import org.testng.annotations.AfterClass;
import org.testng.annotations.BeforeClass;
import org.testng.annotations.DataProvider;
import org.testng.annotations.Test;
```

```
import com.lujiatao.webtestinaction.common.Files;
import com.lujiatao.webtestinaction.jforum.JForum;

public class Account {

    private WebDriver chromeDriver;

    @BeforeClass
    public void init() {
        chromeDriver = new ChromeDriver();
    }

    @SuppressWarnings("unchecked")
    @Test(dataProvider = "loginData")
    public void login(Object[] testData) {
        HashMap<String, String> data = ((ArrayList<HashMap<String, String>>)
            testData[0]).get(0);
        boolean result = JForum.login(chromeDriver, data.get("url"),
            data.get("username"), data.get("password"),
                Boolean.valueOf(data.get("autoLogin")));
        Assert.assertTrue(result);
    }

    @AfterClass
    public void clear() {
        chromeDriver.quit();
    }

    @DataProvider(name = "loginData")
    public Object[] loginData() {
        String dir = System.getProperty("user.dir") + File.separator +
            "src/test/resources/test-data";
        return new Object[] { Files.readExcel(dir, "main.xlsx", "Account") };
    }

}
```

下面对上述代码进行说明。

① 对 init()方法和 clear()方法分别进行初始化和清理操作。

② login()方法为测试用例，该方法接收 Object[]类型的测试数据。login()方法直接调用了业务函数 login()，入参来自名为“loginData”的 DataProvider。

③ loginData()方法为 DataProvider，通过读取 Excel 文件将测试数据提供给测试用例方法 login()。

其次，在 com.lujiatao.webtestinaction.section Package 中创建名为 Section 的 Class，在 Section 中输入以下代码。

```
package com.lujiatao.webtestinaction.section;

import java.io.File;
import java.util.ArrayList;
import java.util.HashMap;

import org.openqa.selenium.WebDriver;
import org.openqa.selenium.chrome.ChromeDriver;
import org.testng.Assert;
import org.testng.annotations.AfterClass;
import org.testng.annotations.BeforeClass;
import org.testng.annotations.DataProvider;
import org.testng.annotations.Test;

import com.lujiatao.webtestinaction.common.Files;
import com.lujiatao.webtestinaction.jforum.JForum;

public class Section {

    private WebDriver chromeDriver;

    @BeforeClass
    public void init() {
        chromeDriver = new ChromeDriver();
    }

    @SuppressWarnings("unchecked")
    @Test(dataProvider = "loginData", priority = 1)
    public void login(Object[] testData) {
        HashMap<String, String> data = ((ArrayList<HashMap<String, String>>)
            testData[0]).get(0);
        boolean result = JForum.login(chromeDriver, data.get("url"),
            data.get("username"), data.get("password"),
                Boolean.valueOf(data.get("autoLogin")));
        Assert.assertTrue(result);
    }

    @SuppressWarnings("unchecked")
    @Test(dataProvider = "addTopicData", priority = 2)
    public void addTopic(Object[] testData) {
        HashMap<String, String> data = ((ArrayList<HashMap<String, String>>)
            testData[0]).get(0);
        Assert.assertTrue(JForum.goSection(chromeDriver, data.get("section")));
        Assert.assertTrue(JForum.goAddTopic(chromeDriver));
        boolean result = JForum.addTopic(chromeDriver, data.get("title"),
            data.get("content"));
        Assert.assertTrue(result);
    }
```

```
    @AfterClass
    public void clear() {
        chromeDriver.quit();
    }

    @DataProvider(name = "loginData")
    public Object[] loginData() {
        String dir = System.getProperty("user.dir") + File.separator +
            "src/test/resources/test-data";
        return new Object[] { Files.readExcel(dir, "main.xlsx", "Account") };
    }

    @DataProvider(name = "addTopicData")
    public Object[] addTopicData() {
        String dir = System.getProperty("user.dir") + File.separator +
            "src/test/resources/test-data";
        return new Object[] { Files.readExcel(dir, "section.xlsx", "Section") };
    }

}
```

下面对上述代码进行说明。

① 将 login()方法的 priority 设置为 1，将 addTopic()方法的 priority 设置为 2，保证先登录再发帖的顺序。

② 在 addTopic()方法中，Assert.assertTrue()方法执行 3 次断言，前两次断言可理解为 Check 点，最后一次断言才是真正的测试用例断言。

③ 使用两个 DataProvider 分别为两条测试用例提供数据。初始化和清理操作与 Account 中保持一致。

9.2.4　测试数据准备

在本节的实战项目中，笔者使用了 Excel 文件存放测试数据。

在 src/test/resources 中分别创建名为 main.xlsx 和 section.xlsx 的文件，并在文件中填写测试数据，分别如图 9-8 和图 9-9 所示。

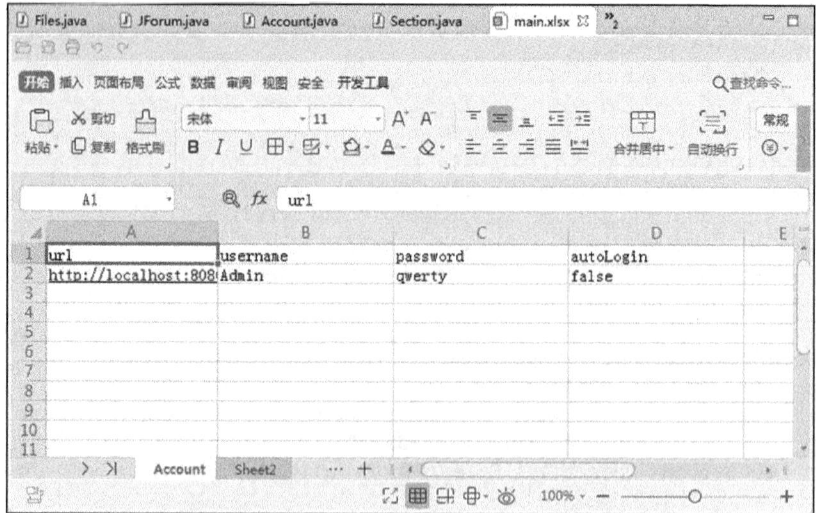

	A	B	C	D
1	url	username	password	autoLogin
2	http://localhost:808	Admin	qwerty	false

图 9-8

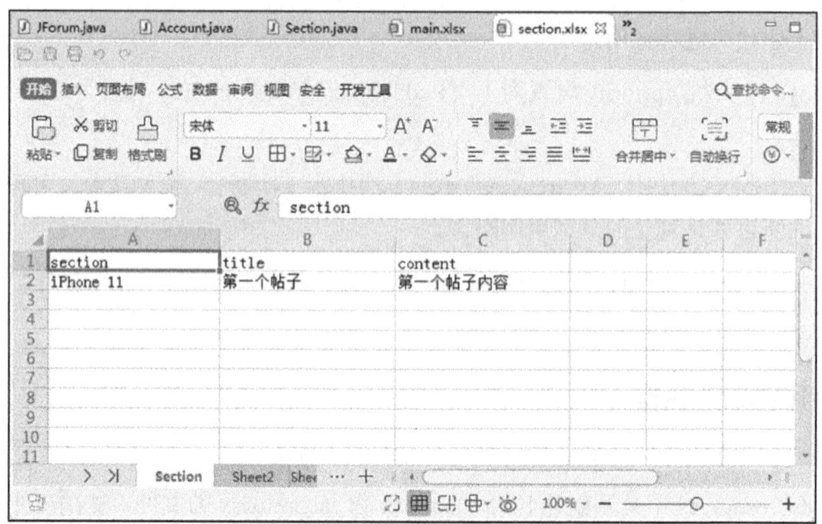

	A	B	C
1	section	title	content
2	iPhone 11	第一个帖子	第一个帖子内容

图 9-9

修改 testng.xml 文件，在<test>标签中新增以下粗体部分内容：

```
<?xml version="1.0" encoding="UTF-8"?>
<!DOCTYPE suite SYSTEM "http://testng.org/testng-1.0.dtd">
<suite name="Suite">
    <test thread-count="5" name="Test">
        <classes>
            <class name="com.lujiatao.webtestinaction.main.Account" />
            <class name="com.lujiatao.webtestinaction.section.Section" />
        </classes>
    </test> <!-- Test -->
```

```
</suite> <!-- Suite -->
```

保存所做的修改，在 testng.xml 上单击右键，从弹出的快捷菜单中选择“Run As → TestNG Suite”选项，查看测试报告，如图 9-10 所示。

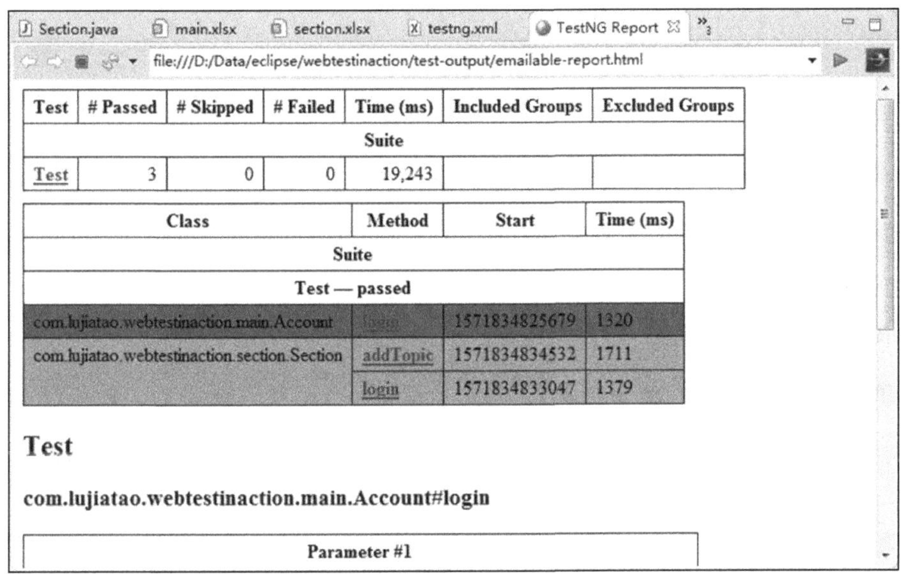

图 9-10

此时查看 JForum 论坛的 iPhone 11 版面，可以看到新发表的帖子，如图 9-11 所示。

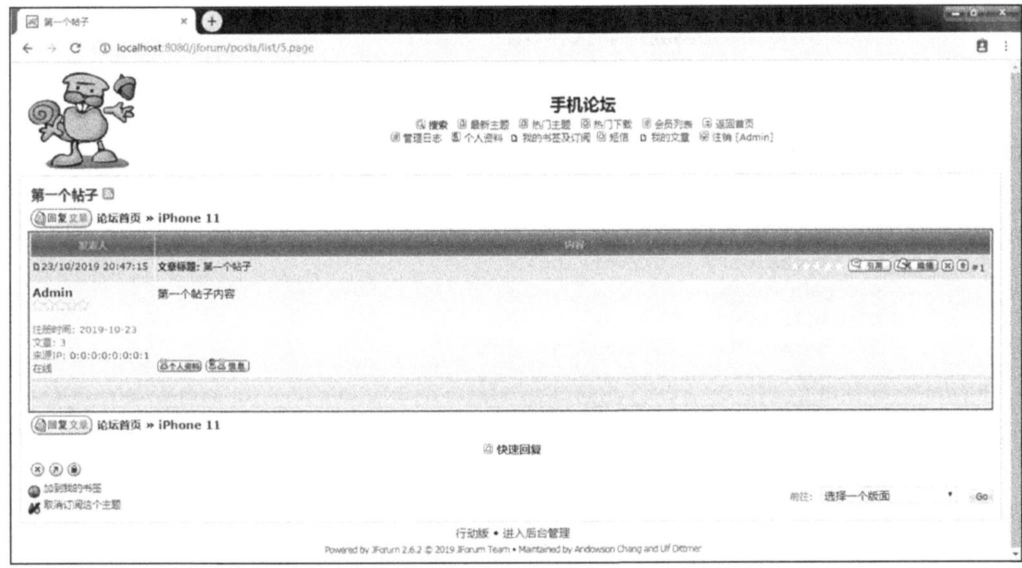

图 9-11

至此，整个 Web 自动化测试实战结束，除用到前面章节介绍的 TestNG 和 Selenium 知识外，着重讲解了自动化测试后实际项目中非常重要的分层和解耦策略。

9.3 Android 自动化测试实战

9.3.1 工程准备

1. 工程创建

创建一个新的 Maven 项目，关键信息填写如下。

Group Id：com.lujiatao

Artifact Id：androidtestinaction

Name：Android Test In Action

读者可根据实际情况填写，不需要和笔者填写的完全一致。

创建完成后，在 pom.xml 文件的<name>标签后输入以下粗体部分内容：

```
<project xmlns="http://maven.apache……"
    xmlns:xsi="http://www.w3……"
    xsi:schemaLocation="http://maven.apache…… http://maven.apache……">
    <modelVersion>4.0.0</modelVersion>

    <groupId>com.lujiatao</groupId>
    <artifactId>androidtestinaction</artifactId>
    <version>0.0.1-SNAPSHOT</version>
    <name>Android Test In Action</name>

    <dependencies>
        <dependency>
            <groupId>io.appium</groupId>
            <artifactId>java-client</artifactId>
            <version>7.1.0</version>
        </dependency>
        <dependency>
            <groupId>org.testng</groupId>
            <artifactId>testng</artifactId>
            <version>6.14.3</version>
            <scope>test</scope>
        </dependency>
    </dependencies>

</project>
```

保存 pom.xml 文件，这时 Maven 会自动下载 Appium、TestNG 及其依赖的其他 jar 包。

当依赖 jar 包下载完成后，在工程（androidtestinaction）上单击右键，从弹出的快捷菜单中选择“TestNG → Convert to TestNG”选项，在工程中生成 testng.xml 文件。

2．构建工程结构

按照图 9-12 所示，新建对应的 Package 和目录。

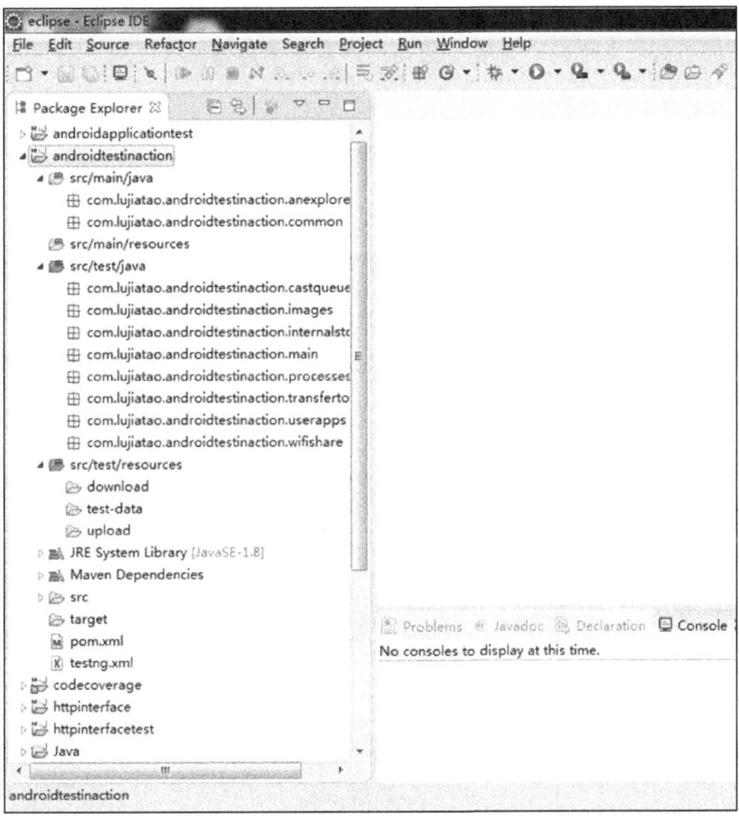

图 9-12

该工程结构符合 9.2.1 节介绍的分层和解耦策略，不再赘述。

9.3.2　Page Object 设计模式

Page Object 设计模式通常用于 Web 自动化测试，但它同样适用于其他类型的 UI 自动化测试。Page Object 设计模式是对页面的操作和元素进行建模，并与具体的自动化测试用例分离，一般分为 3 层。

- Page Object：页面对象层，存放页面对象。
- Business Logic：业务逻辑层，具体的业务逻辑，调用页面对象中的操作组合而成。

- Test Case：测试用例层，具体的测试用例，调用业务逻辑层中的业务逻辑组合而成。另外，测试用例中需要加入断言，以判断测试用例的执行结果。

在使用 Page Object 设计模式实施自动化测试的过程中，需要遵循一定的规则，这些规则的要点如下。

要点 1：public 方法代表页面对象提供的操作。

要点 2：遵循面向对象中的“封装”原则，即尽量不要暴露页面对象的内部。

要点 3：public 方法一般不做断言。

要点 4：public 方法的返回值为其他页面对象。

要点 5：页面对象并非必须代表整个页面。

要点 6：相同的操作、不同的结果被建模为不同的 public 方法。

AnExplorer 作为文件管理器，最重要的功能是对文件的查看、编辑、创建和删除。在本节实战中，将遵循 Page Object 设计模式编写一个创建文件夹的自动化测试用例。各层之间的调用关系如图 9-13 所示。

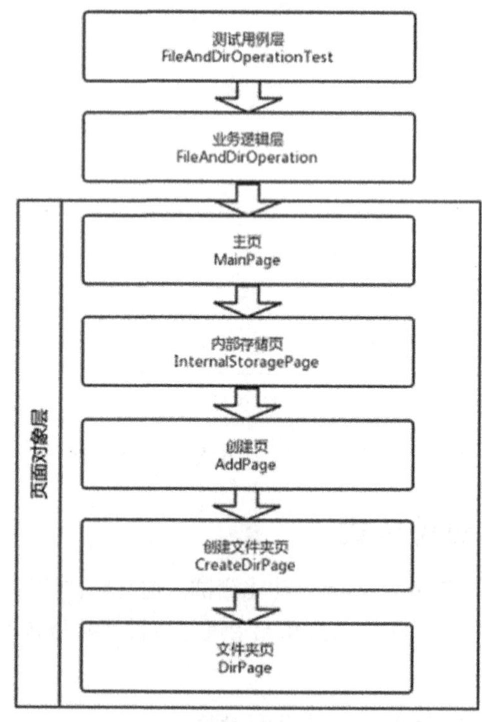

图 9-13

9.3.3　页面对象层封装

1. Page 类封装

Page 类是所有页面对象的基类，其他页面对象均直接或间接继承至 Page 类，该类与具体业务无关。

在 com.lujiatao.androidtestinaction.common Package 中创建名为 Page 的 Class，在 Page 中输入以下代码：

```
package com.lujiatao.androidtestinaction.common;

import org.openqa.selenium.WebDriver;

public class Page {

    private WebDriver driver;

    public Page(WebDriver driver) {
        this.driver = driver;
    }

    public void setDriver(WebDriver driver) {
        this.driver = driver;
    }

    public WebDriver getDriver() {
        return driver;
    }

}
```

上述代码封装了一个 Page 类，构造方法通过传入一个 WebDriver 对象初始化 Page 类的 driver 属性，并构建一个 Page 实例。另外，该类还提供了给 driver 属性赋值和获取的方法。

2. MainPage 类封装

在 com.lujiatao.androidtestinaction.anexplorer Package 中创建 pageobject 子 Package，用于存放 AnExplorer 的页面对象。

创建名为 MainPage 的 Class，在 MainPage 中输入以下代码：

```
package com.lujiatao.androidtestinaction.anexplorer.pageobject;

import com.lujiatao.androidtestinaction.common.Page;

import io.appium.java_client.android.AndroidDriver;
import io.appium.java_client.android.AndroidElement;

public class MainPage extends Page {
```

```
@SuppressWarnings("unchecked")
private AndroidDriver<AndroidElement> driver = (AndroidDriver<AndroidElement>)
    getDriver();
private AndroidElement menu = driver.findElementByAccessibilityId("显示根目录");
private AndroidElement support = driver.findElementByAccessibilityId("Support
    AnExplorer");
private AndroidElement more = driver.findElementByAccessibilityId("更多选项");
private AndroidElement internalStorage = driver.findElementByXPath(
    "//androidx.recyclerview.widget.RecyclerView/android.widget.
        FrameLayout[1]/android.widget.LinearLayout");
private AndroidElement processes = driver.findElementByXPath(
    "//androidx.recyclerview.widget.RecyclerView/android.widget.
        FrameLayout[2]/android.widget.LinearLayout");
private AndroidElement transferToPc = driver.findElementByXPath(
    "//androidx.recyclerview.widget.RecyclerView/android.widget.
        FrameLayout[3]/android.widget.LinearLayout");
private AndroidElement wifiShare = driver.findElementByXPath(
    "//androidx.recyclerview.widget.RecyclerView/android.widget.
        FrameLayout[4]/android.widget.LinearLayout");
private AndroidElement castQueue = driver.findElementByXPath(
    "//androidx.recyclerview.widget.RecyclerView/android.widget.
        FrameLayout[5]/android.widget.LinearLayout");
private AndroidElement userApps = driver.findElementByXPath(
    "//androidx.recyclerview.widget.RecyclerView/android.widget.
        FrameLayout[6]/android.widget.LinearLayout");
private AndroidElement images = driver.findElementByXPath(
    "//androidx.recyclerview.widget.RecyclerView/android.widget.
        FrameLayout[7]/android.widget.LinearLayout");

public MainPage(AndroidDriver<AndroidElement> driver) {
    super(driver);
}

public SidebarPage clickMenu() {
    menu.click();
    return new SidebarPage(driver);
}

public SupportPage clickSupport() {
    support.click();
    return new SupportPage(driver);
}

public MorePage clickMore() {
    more.click();
    return new MorePage(driver);
}

public InternalStoragePage clickInternalStorage() {
```

```
        internalStorage.click();
        return new InternalStoragePage(driver);
    }

    public ProcessesPage clickProcesses() {
        processes.click();
        return new ProcessesPage(driver);
    }

    public TransferToPCPage clickTransferToPc() {
        transferToPc.click();
        return new TransferToPCPage(driver);
    }

    public WiFiSharePage clickWifiShare() {
        wifiShare.click();
        return new WiFiSharePage(driver);
    }

    public CastQueuePage clickCastQueue() {
        castQueue.click();
        return new CastQueuePage(driver);
    }

    public UserAppsPage clickUserApps() {
        userApps.click();
        return new UserAppsPage(driver);
    }

    public ImagesPage clickImages() {
        images.click();
        return new ImagesPage(driver);
    }

}
```

下面对上述代码进行说明。

① 在 MainPage 的构造方法中调用了父类 Page 的构造方法来构造 MainPage。

② 使用 Appium 打开 AnExplorer 主页，如图 9-14 所示。

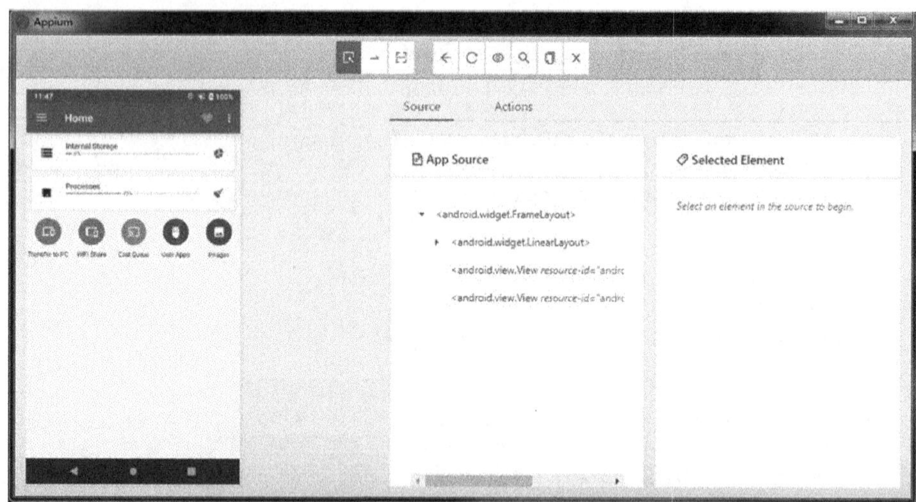

图 9-14

在 AnExplorer 主页中可用的操作有打开侧边栏、支持页面、更多页面、内部存储页面、进程页面、传到 PC、热点分享、投屏、用户应用和图片共计 10 个操作。首先对可操作元素进行定位，然后封装对应方法进行操作，每个方法均返回其他页面对象。

3. InternalStoragePage 类封装

在 com.lujiatao.androidtestinaction.anexplorer.pageobject Package 中创建名为 InternalStoragePage 的 Class，在 InternalStoragePage 中输入以下代码：

```
package com.lujiatao.androidtestinaction.anexplorer.pageobject;

import org.openqa.selenium.WebDriver;

import com.lujiatao.androidtestinaction.common.Page;

import io.appium.java_client.android.AndroidDriver;
import io.appium.java_client.android.AndroidElement;

public class InternalStoragePage extends Page {

    @SuppressWarnings("unchecked")
    private AndroidDriver<AndroidElement> driver = (AndroidDriver<AndroidElement>)
        getDriver();
    private AndroidElement menu = driver.findElementByAccessibilityId("显示根目录");
    private AndroidElement search = driver.findElementByAccessibilityId("搜索");
    private AndroidElement support = driver.findElementByAccessibilityId("Support
        AnExplorer");
    private AndroidElement more = driver.findElementByXPath("//android.widget.
        ImageView[@content-desc='更多选项']");
```

```
    private AndroidElement list = driver.findElementById("dev.dworks.apps.
        anexplorer:id/recyclerview");
    private AndroidElement add = driver.findElementByAccessibilityId("Add");

    public InternalStoragePage(WebDriver driver) {
        super(driver);
    }

    public SidebarPage clickMenu() {
        menu.click();
        return new SidebarPage(driver);
    }

    public SupportPage clickSupport() {
        support.click();
        return new SupportPage(driver);
    }

    public SearchPage clickSearch() {
        search.click();
        return new SearchPage(driver);
    }

    public MorePage clickMore() {
        more.click();
        return new MorePage(driver);
    }

    public DirPage clickList() {
        list.click();
        return new DirPage(driver);
    }

    public AddPage clickAdd() {
        add.click();
        return new AddPage(driver);
    }

}
```

InternalStoragePage 类为内部存储页面，代码整体思路与 MainPage 类的代码类似，不再赘述。

4. AddPage 类封装

在 com.lujiatao.androidtestinaction.anexplorer.pageobject Package 中创建名为 AddPage 的 Class，在 AddPage 中输入以下代码：

```
package com.lujiatao.androidtestinaction.anexplorer.pageobject;

import org.openqa.selenium.WebDriver;
```

```
import com.lujiatao.androidtestinaction.common.Page;

import io.appium.java_client.android.AndroidDriver;
import io.appium.java_client.android.AndroidElement;

public class AddPage extends Page {

    @SuppressWarnings("unchecked")
    private AndroidDriver<AndroidElement> driver = (AndroidDriver<AndroidElement>)
getDriver();
    private AndroidElement menu = driver.findElementByAccessibilityId("显示根目录");
    private AndroidElement search = driver.findElementByAccessibilityId("搜索");
    private AndroidElement support = driver.findElementByAccessibilityId("Support
        AnExplorer");
    private AndroidElement more = driver.findElementByXPath("//android.widget.
        ImageView[@content-desc='更多选项']");
    private AndroidElement list = driver.findElementById("dev.dworks.apps.
        anexplorer:id/recyclerview");
    private AndroidElement createFile = driver.findElementByXPath(
        "//android.widget.LinearLayout[@content-desc='Add']/
            android.widget.LinearLayout/android.widget.LinearLayout[1]");
    private AndroidElement createDir = driver.findElementByXPath(
        "//android.widget.LinearLayout[@content-desc='Add']/android.
            widget.LinearLayout/android.widget.LinearLayout[2]");
    private AndroidElement cancel = driver.findElementById("dev.dworks.apps.
        anexplorer:id/fab");

    public AddPage(WebDriver driver) {
        super(driver);
    }

    public SidebarPage clickMenu() {
        menu.click();
        return new SidebarPage(driver);
    }

    public SupportPage clickSupport() {
        support.click();
        return new SupportPage(driver);
    }

    public SearchPage clickSearch() {
        search.click();
        return new SearchPage(driver);
    }

    public MorePage clickMore() {
        more.click();
        return new MorePage(driver);
```

```
    }

    public InternalStoragePage clickList() {
        list.click();
        return new InternalStoragePage(driver);
    }

    public CreateFilePage clickCreateFile() {
        createFile.click();
        return new CreateFilePage(driver);
    }

    public CreateDirPage clickCreateDir() {
        createDir.click();
        return new CreateDirPage(driver);
    }

    public InternalStoragePage clickCancel() {
        cancel.click();
        return new InternalStoragePage(driver);
    }

}
```

AddPage 类为创建页面，通过单击内部存储页面右下角的“+”图标可进入该页。代码整体思路与 MainPage 类的代码相似，不再赘述。

5. CreateDirPage 类封装

在 com.lujiatao.androidtestinaction.anexplorer.pageobject Package 中创建名为 CreateDirPage 的 Class，在 CreateDirPage 中输入以下代码：

```
package com.lujiatao.androidtestinaction.anexplorer.pageobject;

import org.openqa.selenium.WebDriver;

import com.lujiatao.androidtestinaction.common.Page;

import io.appium.java_client.android.AndroidDriver;
import io.appium.java_client.android.AndroidElement;

public class CreateDirPage extends Page {

    @SuppressWarnings("unchecked")
    private AndroidDriver<AndroidElement> driver =
       (AndroidDriver<AndroidElement>) getDriver();
    private AndroidElement input = driver.findElementById("android:id/text1");
    private AndroidElement cancel = driver.findElementById("android:id/button2");
    private AndroidElement ok = driver.findElementById("android:id/button1");

    public CreateDirPage(WebDriver driver) {
```

```
        super(driver);
    }

    public CreateDirPage inputContent(String content) {
        input.sendKeys(content);
        return this;
    }

    public InternalStoragePage clickCancel() {
        cancel.click();
        return new InternalStoragePage(driver);
    }

    public DirPage clickOk() {
        ok.click();
        return new DirPage(driver);
    }

}
```

CreateDirPage 类为创建文件夹页面，代码整体思路与 MainPage 类的代码类似，不再赘述。

6. DirPage 类封装

在 com.lujiatao.androidtestinaction.anexplorer.pageobject Package 中创建名为 DirPage 的 Class，在 DirPage 中输入以下代码：

```
package com.lujiatao.androidtestinaction.anexplorer.pageobject;

import org.openqa.selenium.WebDriver;

import io.appium.java_client.android.AndroidDriver;
import io.appium.java_client.android.AndroidElement;

public class DirPage extends InternalStoragePage {

    @SuppressWarnings("unchecked")
    private AndroidDriver<AndroidElement> driver =
        (AndroidDriver<AndroidElement>) getDriver();
    private AndroidElement change = driver.findElementById("dev.dworks.apps.a
        nexplorer:id/stack");

    public DirPage(WebDriver driver) {
        super(driver);
    }

    public ChangePage clickChange() {
        change.click();
        return new ChangePage(driver);
    }
```

```
}
```

DirPage 类为文件夹页面，该封装并不是直接继承至 Page 类，因为 DirPage 类的代码与 InternalStoragePage 类的代码高度相似，区别仅仅为文件夹页面可以下拉切换，而内部存储页不可以，如图 9-15 所示。

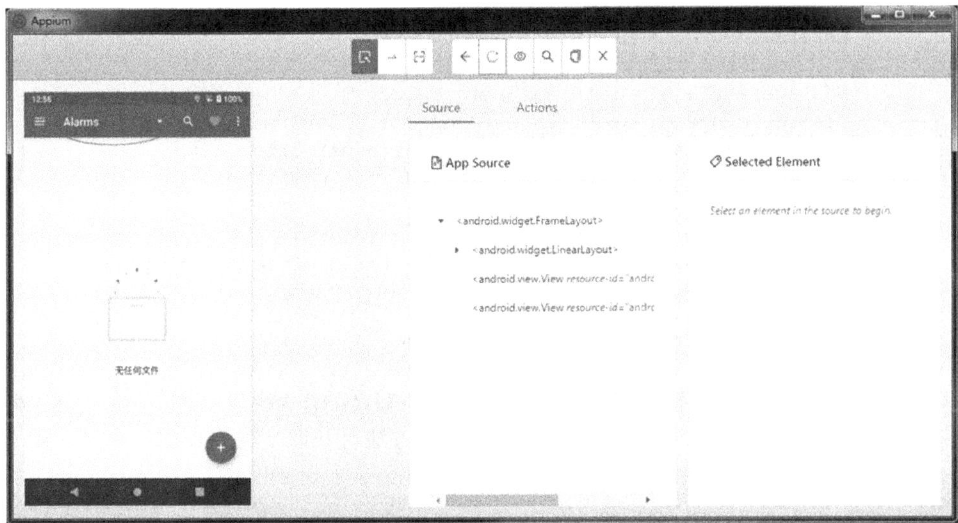

图 9-15

9.3.4　业务逻辑层封装

在 com.lujiatao.androidtestinaction.anexplorer Package 中创建 businesslogic 子 Package，用于存放 AnExplorer 的业务逻辑。

创建名为 FileAndDirOperation 的 Class，在 FileAndDirOperation 中输入以下代码：

```
package com.lujiatao.androidtestinaction.anexplorer.businesslogic;

import com.lujiatao.androidtestinaction.anexplorer.pageobject.AddPage;
import com.lujiatao.androidtestinaction.anexplorer.pageobject.CreateDirPage;
import com.lujiatao.androidtestinaction.anexplorer.pageobject.DirPage;
import
com.lujiatao.androidtestinaction.anexplorer.pageobject.InternalStoragePage;
import com.lujiatao.androidtestinaction.anexplorer.pageobject.MainPage;

import io.appium.java_client.android.AndroidDriver;
import io.appium.java_client.android.AndroidElement;

public class FileAndDirOperation {

    private AndroidDriver<AndroidElement> driver;
    private MainPage mainPage;
    DirPage dirPage;
```

```
    public FileAndDirOperation(AndroidDriver<AndroidElement> driver) {
        this.driver = driver;
        mainPage = new MainPage(driver);
    }

    public void createDir(String name) {
        InternalStoragePage internalStoragePage =
            mainPage.clickInternalStorage();
        AddPage addPage = internalStoragePage.clickAdd();
        CreateDirPage createDirPage = addPage.clickCreateDir();
        dirPage = createDirPage.inputContent(name).clickOk();
    }

    public String getDirTitle() {
        return driver.findElementById("android:id/title").getText();
    }

}
```

出于简化目的，这里只封装了两个方法，即 createDir()方法和 getDirTitle()方法。在 createDir()方法中调用了前面封装的多个页面对象。getDirTitle()方法可获取文件夹页面的标题，用于之后的断言。

9.3.5　自动化测试用例编写

在 src/test/java 的 com.lujiatao.androidtestinaction.internalstorage Package 中创建名为 FileAndDirOperationTest 的 Class，在 FileAndDirOperationTest 中输入以下代码：

```
package com.lujiatao.androidtestinaction.internalstorage;

import java.net.MalformedURLException;
import java.net.URL;

import org.openqa.selenium.remote.DesiredCapabilities;
import org.testng.Assert;
import org.testng.annotations.AfterClass;
import org.testng.annotations.BeforeClass;
import org.testng.annotations.Test;

import com.lujiatao.androidtestinaction.anexplorer.businesslogic.
FileAndDirOperation;

import io.appium.java_client.android.AndroidDriver;
import io.appium.java_client.android.AndroidElement;

public class FileAndDirOperationTest {

    private AndroidDriver<AndroidElement> driver = null;
```

```
    @BeforeClass
    public void init() {
        DesiredCapabilities capabilities = new DesiredCapabilities();
        capabilities.setCapability("platformName", "Android");
        capabilities.setCapability("platformVersion", "9");
        capabilities.setCapability("deviceName", "My Android Device");
        capabilities.setCapability("appPackage", "dev.dworks.apps.anexplorer");
        capabilities.setCapability("appActivity",
            "dev.dworks.apps.anexplorer.DocumentsActivity");
        capabilities.setCapability("noReset", "true");
        try {
            driver = new AndroidDriver<AndroidElement>(new
                URL("http://localhost:4723/wd/hub"), capabilities);
        } catch (MalformedURLException e) {
            e.printStackTrace();
        }
    }

    @Test
    public void createDir() {
        FileAndDirOperation fileAndDirOperation = new
            FileAndDirOperation(driver);
        fileAndDirOperation.createDir("MyDir");
        String expected = "MyDir";
        String actual = fileAndDirOperation.getDirTitle();
        Assert.assertEquals(expected, actual);
    }

    @AfterClass
    public void clear() {
        driver.quit();
    }

}
```

下面对上述代码进行说明。

① 在init()方法和clear()方法中分别进行初始化和清理操作，其中，appPackage和appActivity的获取方法请参阅第 7 章。

② 调用业务逻辑层 FileAndDirOperation 类的 createDir()方法完成文件夹的创建。

③ 调用业务逻辑层 FileAndDirOperation 类的 getDirTitle()方法获取实际值（页面的标题），使用 TestNG 的断言方法完成断言。

修改 testng.xml 文件，在<test>标签中新增以下粗体部分内容：

```
<?xml version="1.0" encoding="UTF-8"?>
<!DOCTYPE suite SYSTEM "http://testng.org/testng-1.0.dtd">
<suite name="Suite">
    <test thread-count="5" name="Test">
```

```
        <classes>
            <class
                name="com.lujiatao.androidtestinaction.internalstorage.
                   FileAndDirOperationTest" />
        </classes>
    </test> <!-- Test -->
</suite> <!-- Suite -->
```

保存所做的修改，在 testng.xml 上单击右键，从弹出的快捷菜单中选择“Run As → TestNG Suite”选项，查看测试报告，如图 9-16 所示。

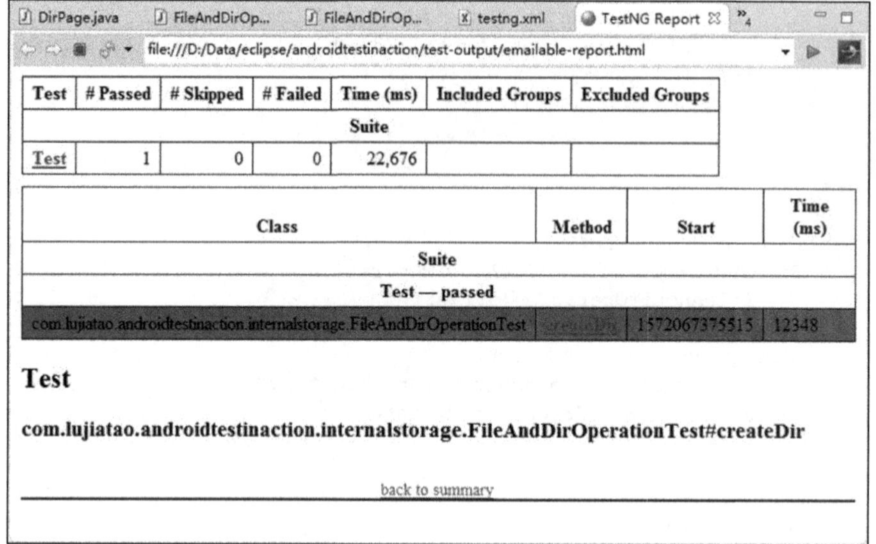

图 9-16

此时在 AnExplorer 中多了一个名为 MyDir 的文件夹，如图 9-17 所示。

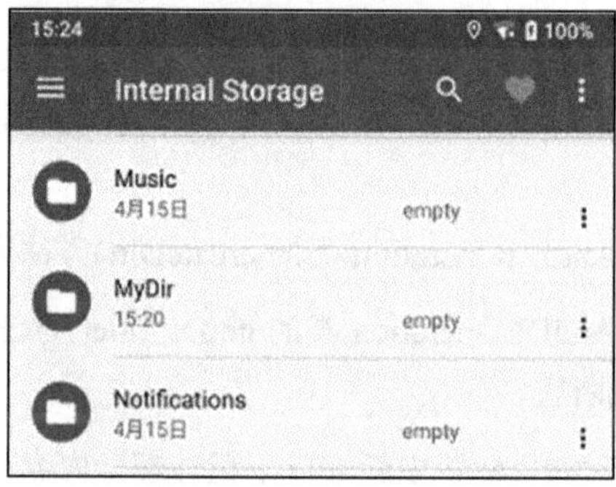

图 9-17

9.4 进一步优化

1. 关于分层策略

如果是大型项目，则可以进行更细致的业务函数分层，比如将业务函数拆分为项目级、模块级和用例级，适用范围如下。

- 项目级业务函数：适用于整个测试项目，可在整个项目中复用。
- 模块级业务函数：适用于整个大模块，可在大模块内部的不同小模块之间复用。
- 用例级业务函数：适用于小模块，可在小模块内部的不同自动化测试用例之间复用。

进行分层之后，整个大型项目的自动化测试用例层次结构更为合理和清晰，规模达到上千甚至上万条自动化测试用例的项目可以考虑采用这种分层策略。

2. 关于解耦策略

在实战项目中，对解耦还简化了配置的解耦。配置解耦适用于有多套自动化测试环境的场景，将不同环境的差异抽离成配置文件。另外，在对测试数据的准备中不同环境也可能会有差异，读者应根据实际项目灵活运用解耦策略。

3. Page Object 设计模式

在 Page Object 设计模式中除页面对象本身外，还可以根据实际需求对页面中的公共部分进行单独抽离，这样可以对代码进行更加有效的复用。

根据上面的策略可以发现在实战项目中有待优化的地方，比如 clickMenu()、clickSupport()和 clickMore()等方法在多个类中均有出现，这时就可以将这些方法单独封装为一个虚拟的（即不是 AnExplorer 上实际存在的）页面对象，用其他页面对象继承它即可。

第10章 持续集成

10.1 持续集成、持续交付和持续部署

持续集成（Continuous Integration，CI）是开发团队经常将本地代码提交到公共分支的一种实践，提交代码后会自动触发新的构建，并通过自动化集成测试进行验证，以确保提交的代码不会影响软件功能。持续集成是持续交付和持续部署的基础。

持续交付（Continuous Delivery，CD）是持续集成的扩展，它属于软件发布过程的自动化，过程如图 10-1 所示。

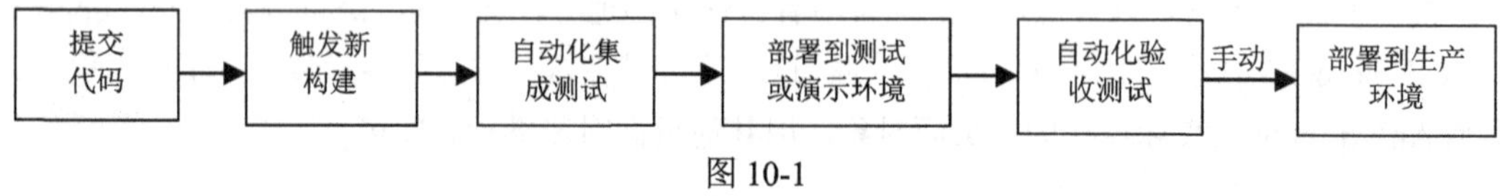

图 10-1

持续部署（Continuous Deployment，CD）是持续交付的扩展。它与持续交付的区别在于：通过自动化验收测试后，持续交付是手动部署到生产环境的，而持续部署是自动化部署到生产环境的，如图 10-2 所示。

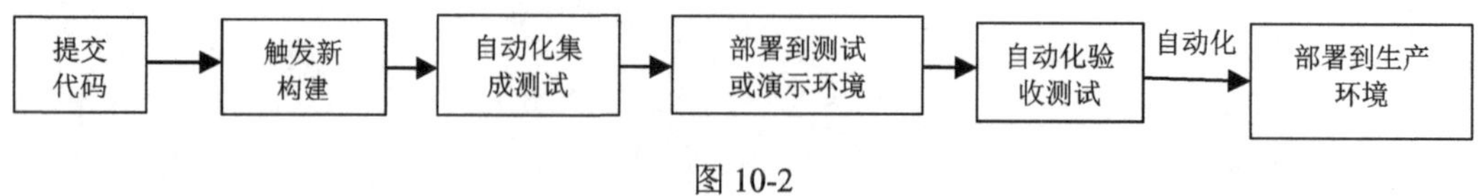

图 10-2

用于持续集成、持续交付和持续部署的中间件有很多，常用的有 Jenkins、TeamCity、Bamboo 和 Hudson 等，本章以 Jenkins 为例，介绍持续集成的相关知识。

10.2 Jenkins 的重要功能简介

由于篇幅所限，本节仅对 Jenkins 的重要功能进行简单介绍。

10.2.1 Jenkins 部署

1．安装 JDK

在即将部署 Jenkins 的服务器上安装 JDK。笔者以本地 Windows 7 电脑作为服务器。

2．部署 Jenkins

下载 Jenkins 程序包，下载后放在服务器上，执行以下命令运行即可：

```
java -jar E:\jenkins.war --httpPort=9090
```

这里将 Jenkins 程序包放在了 E 盘根目录下，读者可根据实际情况替换以上路径。--httpPort 参数指定了端口，如果不指定，则默认端口为 8080。因为前面章节的待测程序大多运行于 8080 端口，所以这里重新指定了端口，避免端口冲突。运行成功，如图 10-3 所示。

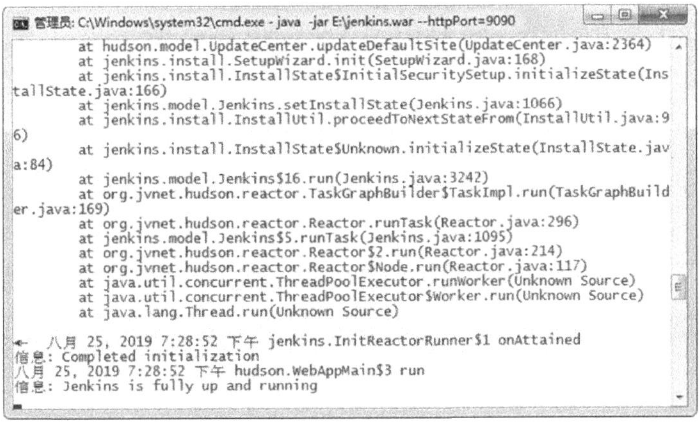

```
管理员: C:\Windows\system32\cmd.exe - java  -jar E:\jenkins.war --httpPort=9090
        at hudson.model.UpdateCenter.updateDefaultSite(UpdateCenter.java:2364)
        at jenkins.install.SetupWizard.init(SetupWizard.java:168)
        at jenkins.install.InstallState$InitialSecuritySetup.initializeState(Ins
tallState.java:166)
        at jenkins.model.Jenkins.setInstallState(Jenkins.java:1066)
        at jenkins.install.InstallUtil.proceedToNextStateFrom(InstallUtil.java:9
6)
        at jenkins.install.InstallState$Unknown.initializeState(InstallState.jav
a:84)
        at jenkins.model.Jenkins$16.run(Jenkins.java:3242)
        at org.jvnet.hudson.reactor.TaskGraphBuilder$TaskImpl.run(TaskGraphBuild
er.java:169)
        at org.jvnet.hudson.reactor.Reactor.runTask(Reactor.java:296)
        at jenkins.model.Jenkins$5.runTask(Jenkins.java:1095)
        at org.jvnet.hudson.reactor.Reactor$2.run(Reactor.java:214)
        at org.jvnet.hudson.reactor.Reactor$Node.run(Reactor.java:117)
        at java.util.concurrent.ThreadPoolExecutor.runWorker(Unknown Source)
        at java.util.concurrent.ThreadPoolExecutor$Worker.run(Unknown Source)
        at java.lang.Thread.run(Unknown Source)

←  八月 25, 2019 7:28:52 下午 jenkins.InitReactorRunner$1 onAttained
信息: Completed initialization
八月 25, 2019 7:28:52 下午 hudson.WebAppMain$3 run
信息: Jenkins is fully up and running
```

图 10-3

细心的读者可以发现，虽然运行成功了，但是 CMD 窗口有报错信息，报错信息的开头为“connect timed out”，意思是连接超时。由于 Jenkins 的服务器在国外，且默认使用 HTTPS 传输，因此会出现连接超时的情况。进入 C:\Users\lujiatao\.jenkins 目录（其中“lujiatao”为笔者本地电脑目录，读者应根据实际情况替换），编辑 hudson.model.UpdateCenter.xml 文件，将<url>标签内的地址 https 修改为 http，保存后重新启动 Jenkins，报错信息就会消失。打开浏览器，访问 http://localhost:9090/，出现“解锁 Jenkins”页面，说明 Jenkins 服务器运行正常，如图 10-4 所示。

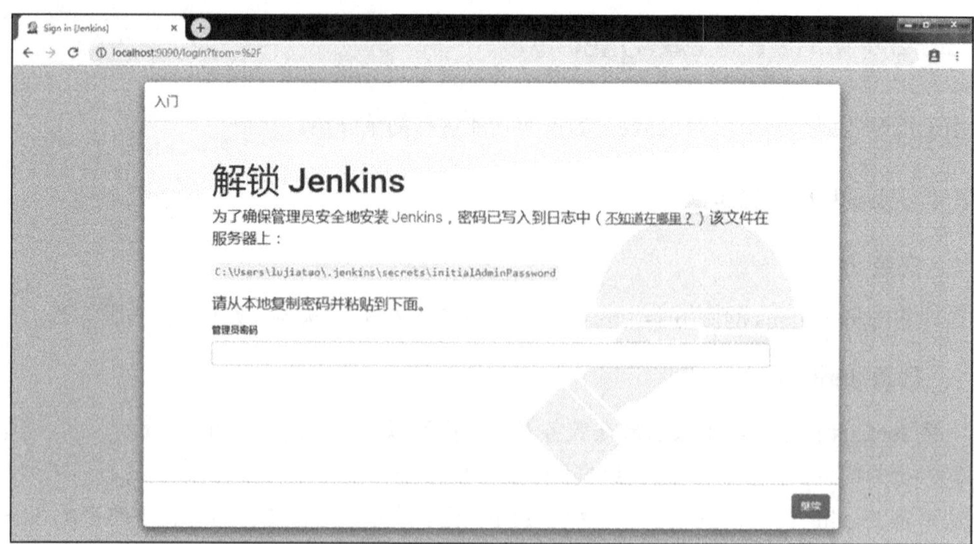

图 10-4

按照页面提示解锁 Jenkins，安装推荐的插件，安装插件过程较长，请耐心等待。安装完成后会提示创建第一个管理员用户。创建完成后，进入“实例配置”页面，其实就是配置 Jenkins 的 URL，这里保持默认就行。

最后会提示“Jenkins 已就绪！”。

重启 Jenkins，刷新空白页面，进入登录页面，登录后即可进入 Jenkins 首页，如图 10-5 所示。

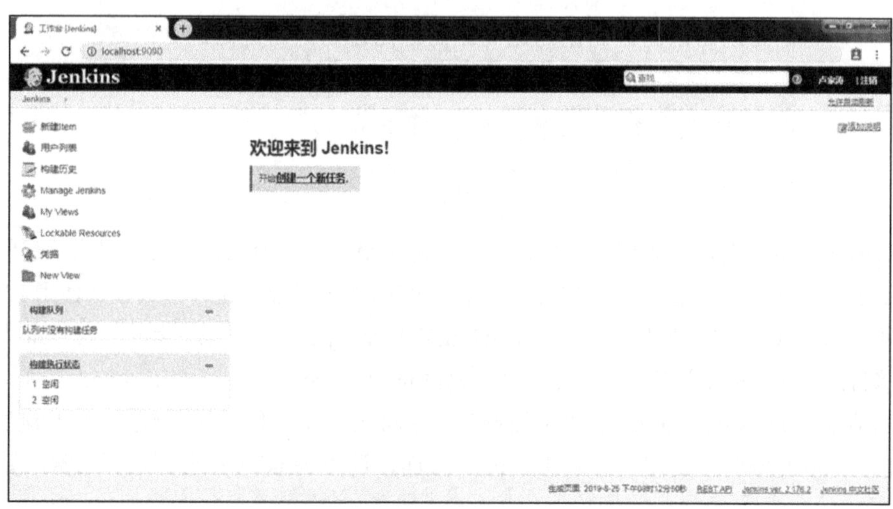

图 10-5

10.2.2 任务管理

Jenkins 最核心的功能是任务，所有的构建触发都是基于任务的。在 Jenkins 首页单击“新建 Item”或“创建一个新任务”，可以进入创建任务页面。默认一共有 6 种任务（可以通过安装 Jenkins 插件增加其他类型的任务），如图 10-6 所示。

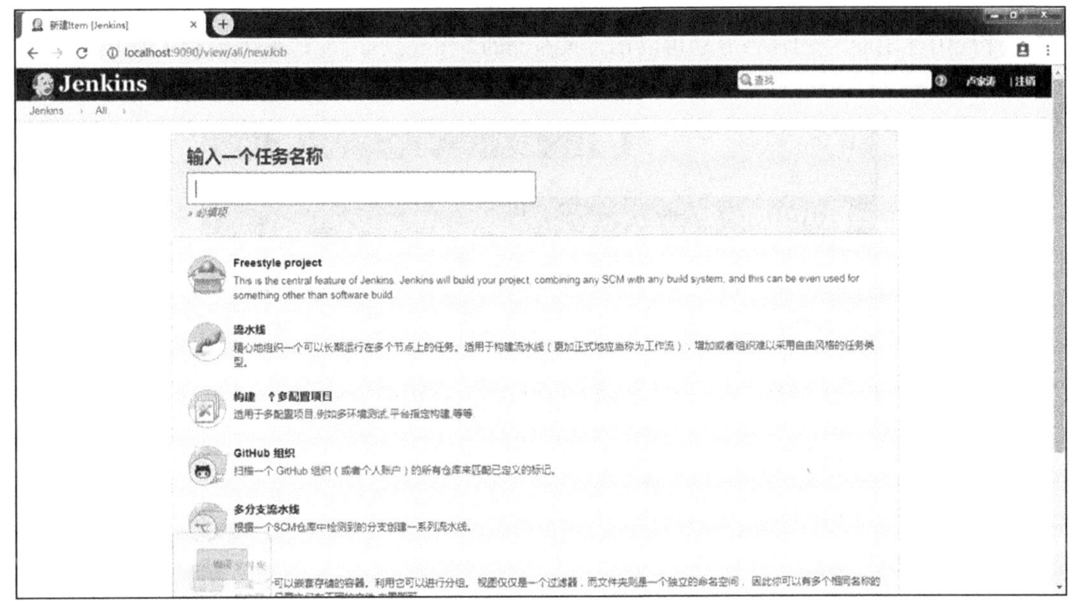

图 10-6

① Freestyle project：自由风格项目，这是 Jenkins 的核心特性，将任何 SCM（Software Configuration Management，软件配置管理）系统与任何构建系统相结合，这种任务还可以用于非软件的构建。

② 流水线：流水线任务可以运行在多个节点上，这种任务通常需要执行多个步骤，比如编译、静态检查、单元测试、部署、自动化接口测试和自动化界面测试等。这种任务如果采用自由风格项目，则往往很难完成。

③ 构建一个多配置项目：适用于多配置项目，这里的多配置包括多套环境、多个节点等。

④ GitHub 组织：这种任务是专门为 GitHub 设计的，它可以扫描一个 GitHub 账号下的所有仓库。

⑤ 多分支流水线：这是流水线任务的“升级版”，代码仓库往往有多个分支，使用多分支流水线可以为每个分支创建一个流水线。

⑥ 文件夹：当 Jenkins 的任务过多时，我们可以使用视图对任务进行分类。但视图仅仅是

一个过滤器，在不同视图中不能包含相同名称的任务。创建文件夹可以真正做到命名空间的隔离，在不同文件夹中可以创建相同名称的任务。

10.2.3　构建管理

任务创建完成后就可以构建任务了，Jenkins 默认可以同时并行构建 2 个任务，超过 2 个将排队，构建中和排队中的任务分别可以在首页左侧的“构建执行状态”和“构建队列中”查看，如图 10-7 所示。

图 10-7

在构建频率较低时，默认的 2 个任务并行构建是可以满足需求的。当构建频率较高时则不适用，这时可以修改并行构建任务的数量。单击“Manage Jenkins → Configure System”进入配置页面，然后修改执行者数量即可，如图 10-8 所示。

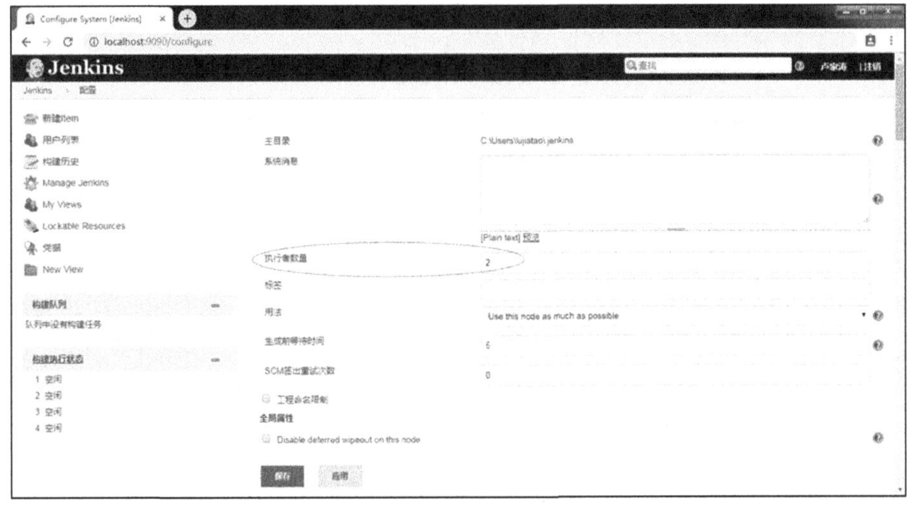

图 10-8

并行构建的任务数量不能设置太多，否则当服务器资源不够时，Jenkins 将停止提供服务，需要重启 Jenkins 才能解决。

10.2.4　节点管理

当一个 Jenkins 节点无法满足频繁的构建需求时，我们可以配置多个节点，默认节点为 master，即主节点，其他节点为 slave，即从节点。单击“Manage Jenkins → Manage Nodes”进入 Nodes 页面，再单击左侧的“新建节点”即可进入新建节点页面，如图 10-9 所示。

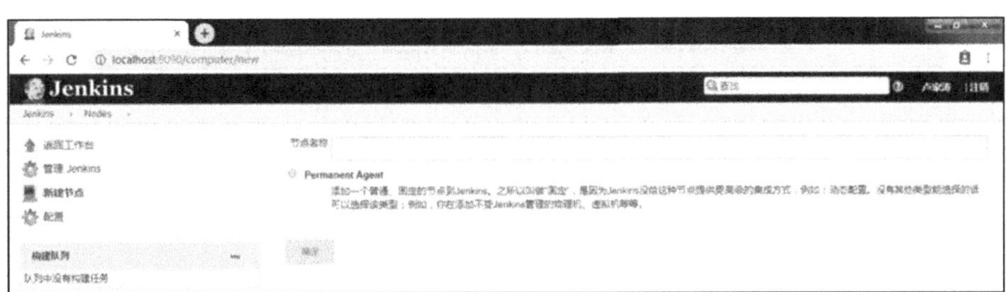

图 10-9

当配置好从节点后，后续就可以在从节点上构建任务，使用这种方法可以大大提升对 Jenkins 的并行构建能力，因为从节点是可以配置多个的。

10.2.5　插件管理

Jenkins 之所以如此受欢迎，有一个很大的原因在于它拥有非常多的第三方插件。单击“Manage Jenkins → Manage Plugins”，进入插件管理页面，切换到“可选插件”Tab 页，可以看到有很多可供选择的插件，如图 10-10 所示。

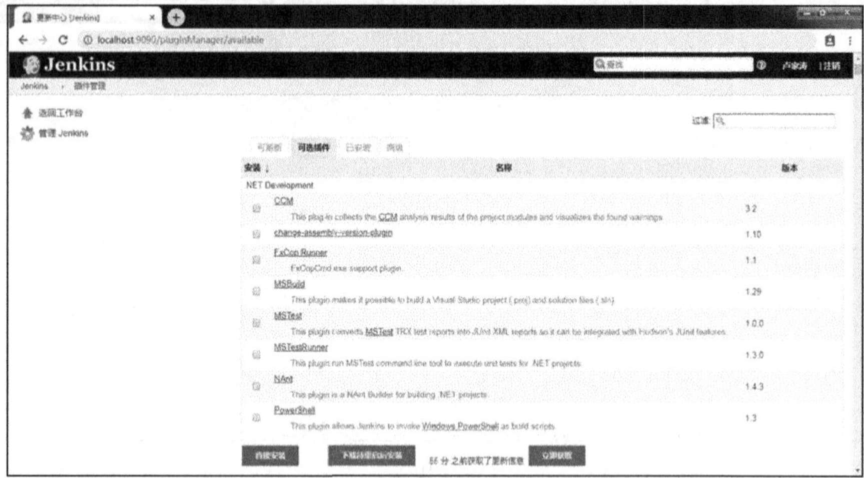

图 10-10

在插件管理页面可以对插件进行安装、升级和卸载。这里重点说明一下安装插件，安装插件分两种方式，即在线安装和离线安装。

在线安装即在图 10-10 所示页面勾选待安装的插件后，单击“直接安装”或“下载待重启后安装”即可安装插件。

离线安装为先获取插件的安装包（hpi 格式），切换到“高级”Tab 页，在“上传插件”中上传插件的安装包，如图 10-11 所示。

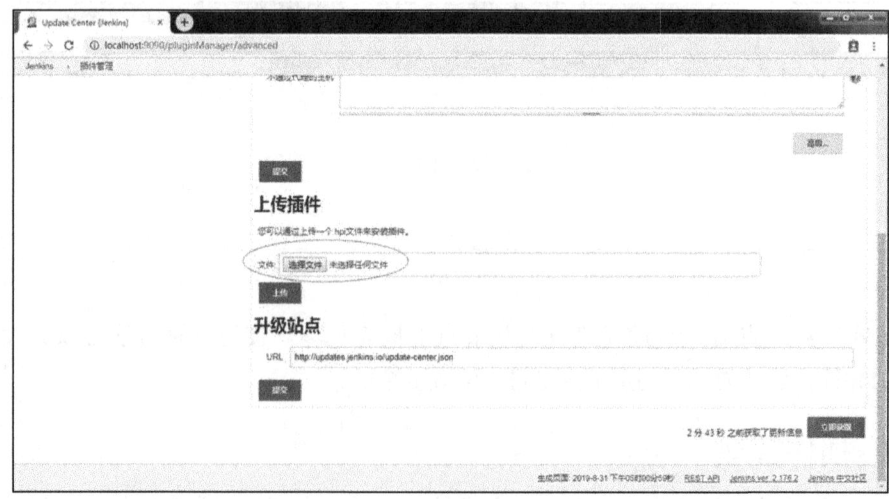

图 10-11

10.2.6　用户管理

Jenkins 有完善的用户管理功能，并且可以对不同的用户设置权限。单击"Manage Jenkins → Manage Users"，进入用户列表页面，如图 10-12 所示。

图 10-12

单击左侧"新建用户"选项，即可新建一个用户。

Jenkins 默认是不允许用户注册的，只能通过上述方法手动添加用户。当然，我们可以将注册功能开放出来，单击"Manage Jenkins → Configure Global Security"，进入 Configure Global Security 页面，勾选"允许用户注册"即可，如图 10-13 所示。

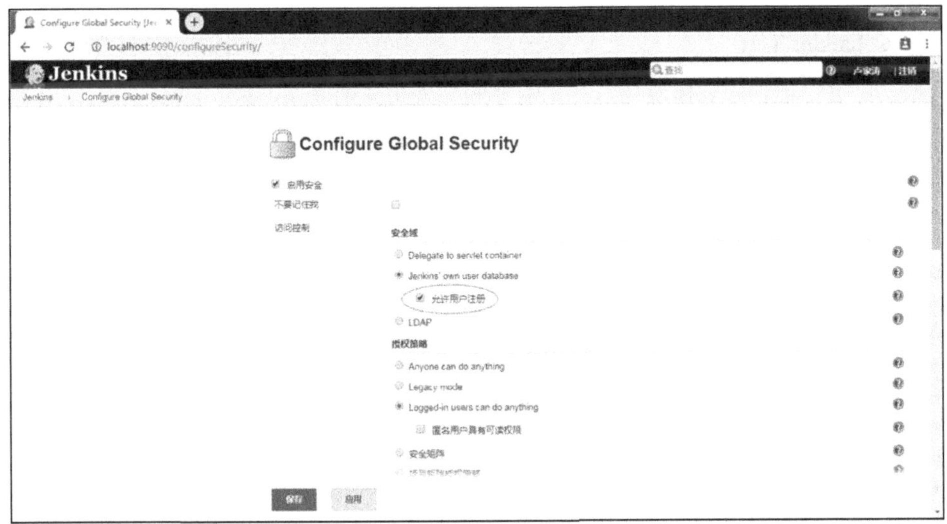

图 10-13

保存后退出登录，可以看到登录页面多了用户注册的入口，如图 10-14 所示。

图 10-14

10.3 TestNG 集成到 Jenkins

10.3.1 TestNG 工程创建

创建一个新的 Maven 项目，关键信息填写如下。

Group Id：com.lujiatao

Artifact Id：jenkinstest

Name：Jenkins Test

当然，读者可根据实际情况填写，不需要和笔者填写的完全一致。

创建完成后，在 pom.xml 文件的<name>标签后输入以下粗体部分内容：

```
<project xmlns="http://maven.apache……"
    xmlns:xsi="http://www.w3……"
    xsi:schemaLocation="http://maven.apache…… http://maven.apache.……">
    <modelVersion>4.0.0</modelVersion>

    <groupId>com.lujiatao</groupId>
    <artifactId>jenkinstest</artifactId>
    <version>0.0.1-SNAPSHOT</version>
    <name>Jenkins Test</name>

    <dependencies>
        <dependency>
            <groupId>org.testng</groupId>
            <artifactId>testng</artifactId>
            <version>6.14.3</version>
            <scope>test</scope>
        </dependency>
    </dependencies>

    <build>
```

```
        <plugins>
            <plugin>
                <groupId>org.apache.maven.plugins</groupId>
                <artifactId>maven-surefire-plugin</artifactId>
                <version>2.22.2</version>
                <configuration>
                    <suiteXmlFiles>
                        <suiteXmlFile>testng.xml</suiteXmlFile>
                    </suiteXmlFiles>
                </configuration>
            </plugin>
        </plugins>
    </build>

</project>
```

保存 pom.xml 文件，这时 Maven 会自动下载 TestNG 及其依赖的其他 jar 包。maven-surefire-plugin 是 Maven 的测试插件，这里作为 Jenkins 构建时执行测试用例，其中，<suiteXmlFile>标签中指定了测试用例的 XML 配置文件。

依赖 jar 包下载完成后，在工程（jenkinstest）上用鼠标右击，从弹出的快捷菜单中选择"TestNG → Convert to TestNG" 选项，在工程中生成 testng.xml 文件。

接下来在 src/test/java 中创建名为 com.lujiatao.jenkinstest 的 Package，以及名为 JenkinsTest 的 Class，在 JenkinsTest 中输入以下代码。

```
package com.lujiatao.jenkinstest;

import org.testng.Assert;
import org.testng.annotations.Test;

public class JenkinsTest {

    @Test
    public void testCase1() {
        Assert.assertEquals(100, 100);
    }

    @Test
    public void testCase2() {
        Assert.assertEquals("100", "100");
    }

    @Test
    public void testCase3() {
        Assert.assertEquals(100, 99);
    }

}
```

修改 testng.xml 文件，在<test>标签中新增以下粗体部分内容：

```
<?xml version="1.0" encoding="UTF-8"?>
<!DOCTYPE suite SYSTEM "http://testng.org/testng-1.0.dtd">
<suite name="Suite">
    <test thread-count="5" name="Test">
        <classes>
            <class name="com.lujiatao.jenkinstest.JenkinsTest" />
        </classes>
    </test> <!-- Test -->
</suite> <!-- Suite -->
```

保存所做的修改，在 testng.xml 上用鼠标右击，从弹出的快捷菜单中选择“Run As → TestNG Suite”选项，查看测试报告，如图 10-15 所示。

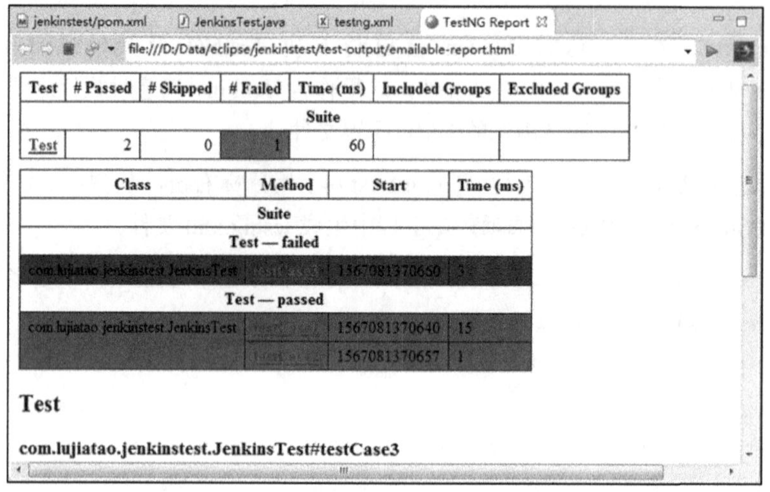

图 10-15

由于 100 和 99 不相等，因此 testCase3 执行失败，执行结果符合预期。

10.3.2 SVN 部署及使用

在实际项目中，一般使用 SVN 或 Git 作为代码仓库，这里以 SVN 为例，服务器使用笔者本地的 Windows 7 电脑。

1. SVN 部署

SVN 分为服务端和客户端，属于典型的 C/S 架构。服务端一般使用 VisualSVN Server，客户端一般使用 TortoiseSVN。

（1）安装 VisualSVN Server

从官网下载 VisualSVN-Server-4.0.3-x64.msi 文件并双击，按照安装向导安装即可。安装路径、仓库和备份默认都在 C 盘，可以手动更改，笔者将它们都安装到了 D 盘。

（2）安装 TortoiseSVN

从官网下载 TortoiseSVN-1.12.2.28653-x64-svn-1.12.2.msi 文件并双击，按照安装向导安装即可。安装完成后，在任意目录用鼠标右击即可看到 SVN 的相关选项，如图 10-16 所示。

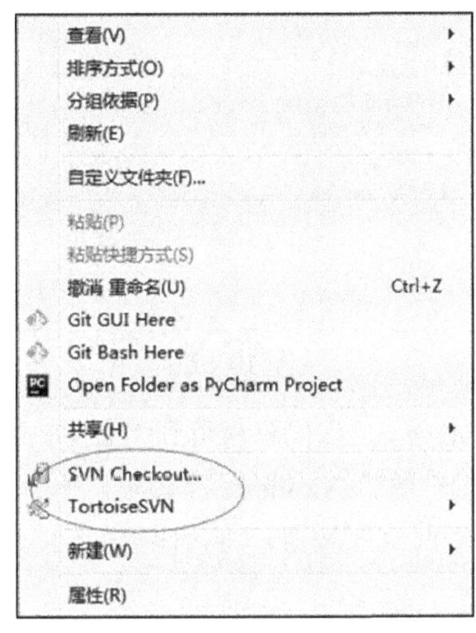

图 10-16

2. SVN 使用

（1）创建代码仓库

在上传代码之前需要先创建代码仓库，否则代码无法提交到 SVN 服务器。

在 Visual SVN Sener 首页中的 Repositories 上用鼠标右击，从弹出的快捷菜单中选择“Create New Repository...”选项，按照创建向导创建即可。仓库名称为 TestNG（读者可随意命名）。

（2）创建 SVN 用户

在 Visual SVN Sener 首页的 Users 上用鼠标右击，从弹出的快捷菜单中选择“Create User...”选项，输入用户名和密码即可创建用户。

（3）上传代码

进入 D 盘根目录，用鼠标右击，从弹出的快捷菜单中选择“SVN Checkout...”选项，在 Checkout 对话框中的 URL of repository 下输入 https://localhost/svn/TestNG。

单击“OK”按钮，此时会弹出认证对话框，单击“Accept the certificate permanently”选项，输入用户名和密码即可将代码仓库 TestNG 签出到本地。当然，在这个示例中，SVN 服务器和客户端在同一台机器上，而在实际项目中，SVN 服务器是在远端的。现在可以看见 D 盘根目录多了一个 TestNG 文件夹，文件夹上有一个钩，如图 10-17 所示。

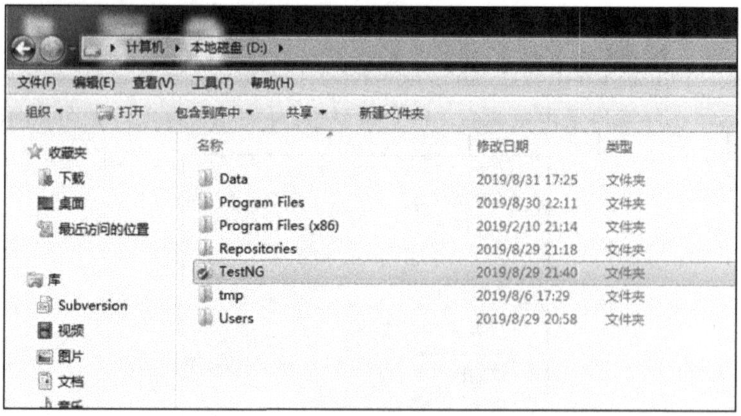

图 10-17

接下来把 10.3.1 节创建的 Maven 工程代码拷贝到 TestNG 文件夹下，在 TestNG 文件夹上用鼠标右击，从弹出的快捷单中选择“SVN Commit...”选项，在 Tortoise SVN 对话中，单击“All”选项，勾选所有文件，然后单击“OK”按钮，将代码提交到 SVN 服务器上。

10.3.3 JDK 和 Maven 配置

在 Maven 官方网站下载并解压缩 Maven 到 D:\Program Files 目录。

单击 Jenkins 的“Manage Jenkins → Global Tool Configuration”，进入 Global Tool Configuration 页面，在 JDK 安装中新增 JDK，在 Maven 安装中新增 Maven，分别如图 10-18 和图 10-19 所示。

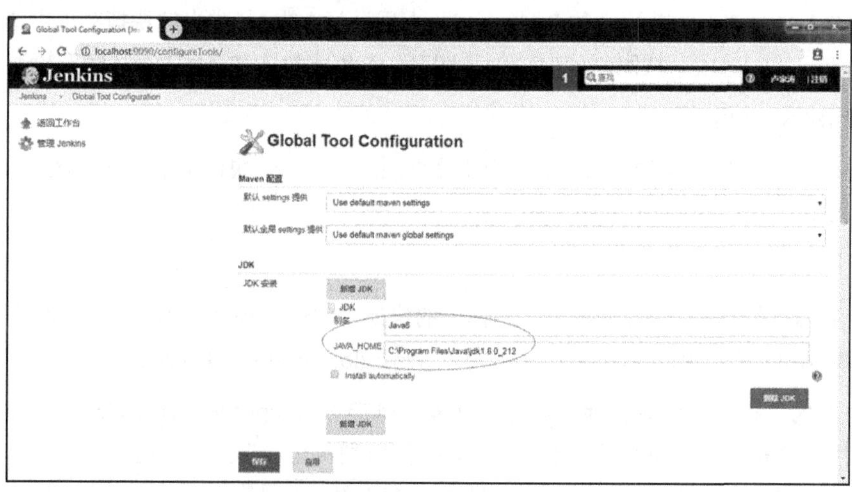

图 10-18

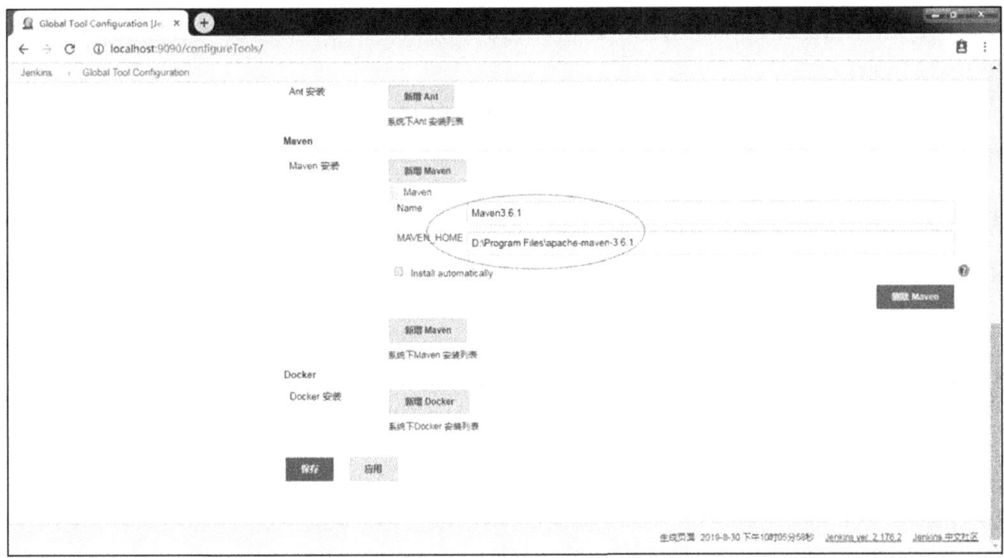

图 10-19

10.3.4　把 TestNG 集成到 Jenkins

1．创建 Jenkins 任务

在 Jenkins 任务创建之前，先安装 TestNG Results 插件，再创建一个名为 TestNGToJenkins 的自由风格项目。在“源码管理”中配置代码路径和认证信息，如图 10-20 所示。

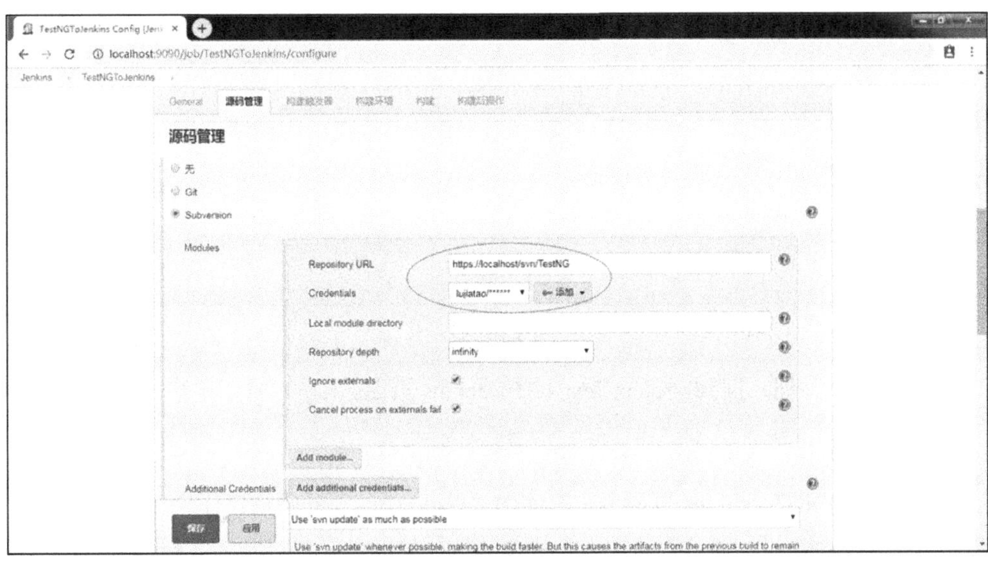

图 10-20

在“构建”中配置 Maven 命令，在“构建后操作”中配置 TestNG 结果，分别如图 10-21 和图 10-22 所示。

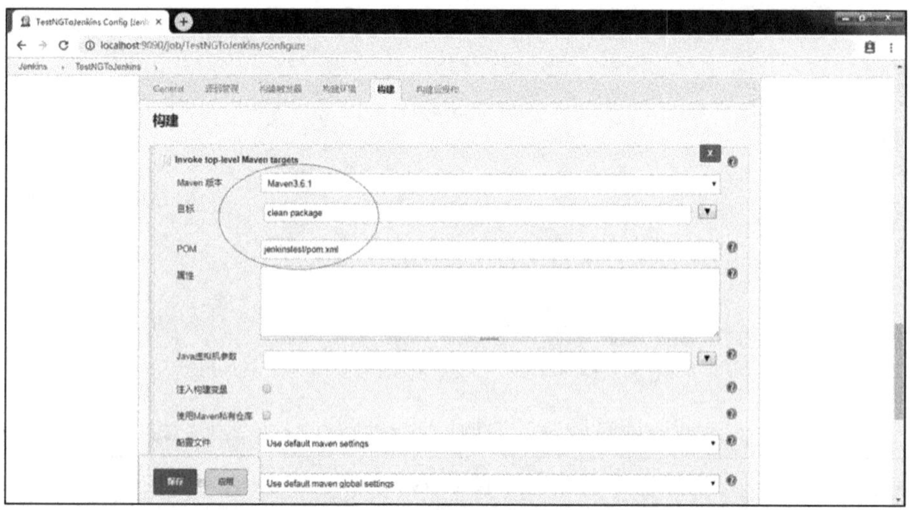

图 10-21

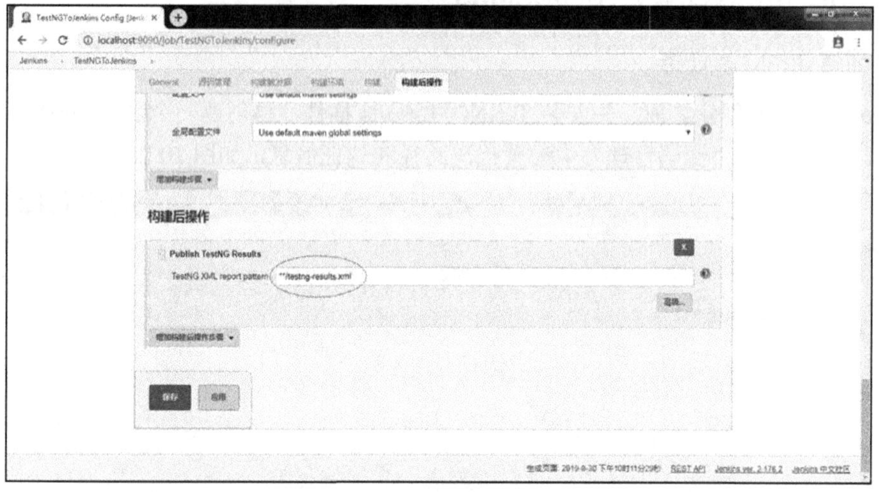

图 10-22

单击“保存”按钮，回到 Project TestNGToJenkins 页面。

2．手动构建

在 Project TestNGToJenkins 页面单击左侧的“Build Now”开始构建任务。构建完成后单击“#3 → TestNG Results”（3 代表构建号，需根据实际情况而定），即可查看 TestNG 结果，如图 10-23 所示。

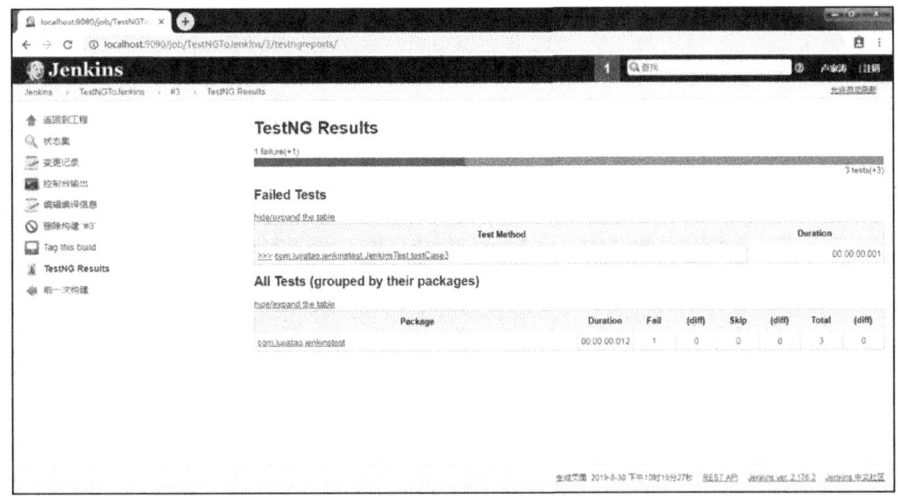

图 10-23

可以看出，通过 Jenkins 执行的 TestNG 自动化用例与在 Eclipse 中执行的结果一致。

3．定时触发构建

在实际项目中，经常会用到 Daily Building，即每日构建，通过每日构建可以及时反馈代码质量。一个实际场景是每日定时从 Dev 分支拉取代码并部署，通过每日构建定时触发自动化用例的 Jenkins 任务来执行测试，这样会及时发现代码缺陷。

定时任务配置需要用到"构建触发器"中的 Build periodically，图 10-24 表示每日凌晨 2:00 构建一次。

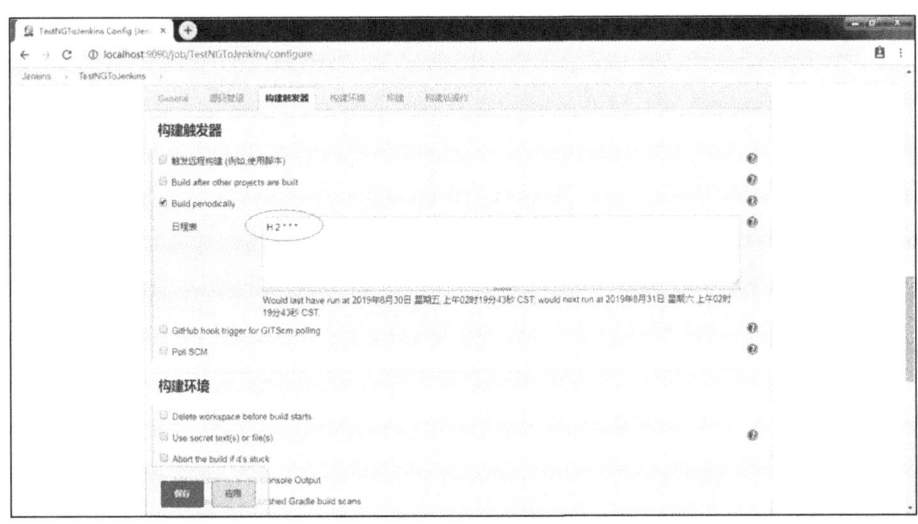

图 10-24

4. 事件触发构建

与定时构建相比，还有一种更为快速反馈代码质量的方式，即根据事件触发构建，这种触发可以使用 SVN 的 post-commit 或 Git 的 Webhook，由于篇幅所限，这里不详细介绍。下面介绍一种“妥协”方案，即通过 Poll SCM 轮询检测代码是否变更，如果变更即触发构建，配置方式如图 10-25 所示。

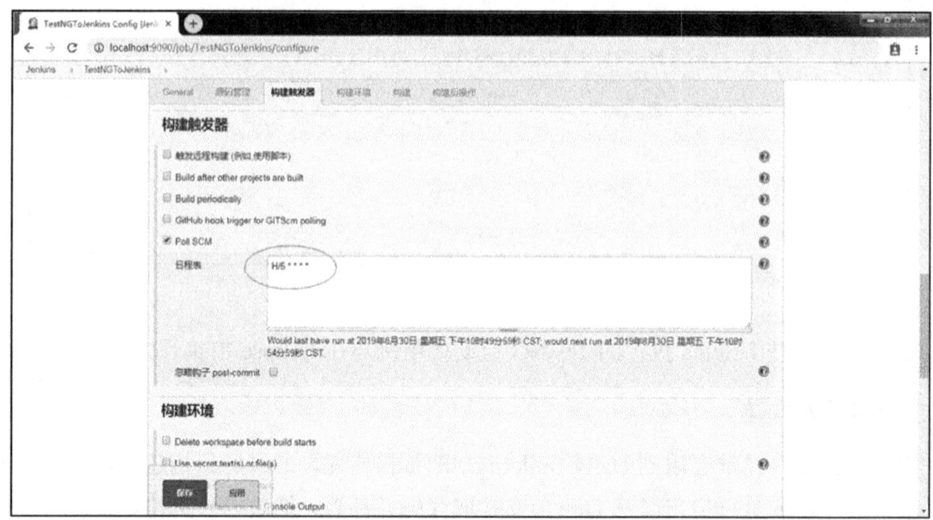

图 10-25

这里配置为每 5 分钟就查看一次代码仓库的代码是否发生变更，如果有变更，则触发构建。

第11章 Mock测试和代码覆盖率

Mock 有“模仿”“虚设”之意，Mock 测试的目的在于模拟依赖方，从而使 Mock 测试只关心测试对象自身的业务逻辑实现，认为依赖方可靠或暂时不对依赖方进行测试。

11.1 单元 Mock 测试

11.1.1 单元 Mock 测试简介

在第 3 章的 Controller 层单元测试中用到了 Spring 自带的 Mock 测试工具 MockMvc，其实有许多专业的 Mock 测试工具，比如 EasyMock、Mockito、PowerMock 和 JMockit 等。其中用得最多的是 Mockito，而功能最强大的是 JMockit。有兴趣的读者可以访问它们的官方网站了解更多信息。

11.1.2 Mockito 用法

学习使用 Mockito 之前，先打开 Eclipse，创建一个新的 Maven 项目，关键信息填写如下。

Group Id：com.lujiatao

Artifact Id：mocktest

Name：Mock Test

当然，读者可根据实际情况填写，不需要和笔者填写的完全一致。

创建完成后，在 pom.xml 文件的<name>标签后输入以下粗体部分内容：

```
<project xmlns="http://maven.apache……"
    xmlns:xsi="http://www.w3……"
    xsi:schemaLocation="http://maven.apache…… http://maven.apache……">
```

```
    <modelVersion>4.0.0</modelVersion>

    <groupId>com.lujiatao</groupId>
    <artifactId>mocktest</artifactId>
    <version>0.0.1-SNAPSHOT</version>
    <name>Mock Test</name>

    <dependencies>
        <dependency>
            <groupId>org.mockito</groupId>
            <artifactId>mockito-core</artifactId>
            <version>3.0.0</version>
            <scope>test</scope>
        </dependency>
        <dependency>
            <groupId>org.testng</groupId>
            <artifactId>testng</artifactId>
            <version>6.14.3</version>
            <scope>test</scope>
        </dependency>
    </dependencies>

</project>
```

保存 pom.xml 文件，这时 Maven 会自动下载 Mockito、TestNG，以及它们依赖的其他 jar 包。

依赖 jar 包下载完成后，在工程(mocktest)上用鼠标右击，从弹出的快捷菜单中选择“TestNG → Convert to TestNG”选项，在工程中生成 testng.xml 文件。

首先创建被测类。在 src/main/java 中创建名为 com.lujiatao.mocktest 的 Package 及名为 Work 的 Class，在 Work 中输入以下代码。

```
package com.lujiatao.mocktest;

public class Work {

    private Weather weather;

    public int getWorkTime() {
        String result = weather.getWeather();
        switch (result) {
        case "晴":
            return 6;
        case "多云":
            return 7;
        case "阴":
            return 8;
        default:
            return 5;
```

```
        }
    }

}
```

被测类中有一个方法为 getWorkTime()，该方法调用了依赖类 Weather 中的 getWeather()方法，可根据不同的天气决定不同的工作时长（假设工作时长的单位为小时）。

然后创建依赖类。在 com.lujiatao.mocktest 中创建名为 Weather 的 Class，在 Weather 中输入以下代码。

```
package com.lujiatao.mocktest;

public class Weather {

    public String getWeather() {
        double tmp = Math.random();
        if (tmp < 0.25) {
            return "晴";
        } else if (0.25 <= tmp && tmp < 0.5) {
            return "多云";
        } else if (0.5 <= tmp && tmp < 0.75) {
            return "阴";
        } else {
            return "雨";
        }
    }

}
```

依赖类是指被测类所依赖的类，该类提供了一个 getWeather()方法，方法中会随机返回“晴”、“多云”、“阴”和“雨”4 种天气。

最后创建测试类。在 src/test/java 中创建名为 com.lujiatao.mocktest 的 Package 及名为 WorkTest 的 Class，在 WorkTest 中输入以下代码。

```
package com.lujiatao.mocktest;

import org.mockito.InjectMocks;
import org.mockito.Mock;
import org.mockito.Mockito;
import org.mockito.MockitoAnnotations;
import org.testng.Assert;
import org.testng.annotations.BeforeClass;
import org.testng.annotations.Test;

public class WorkTest {

    @InjectMocks
    private Work work;
```

```
    @Mock
    private Weather weather;

    @BeforeClass
    public void init() {
        MockitoAnnotations.initMocks(this);
    }

    @Test
    public void testCase1() {
        Mockito.when(weather.getWeather()).thenReturn("晴");
        int actual = work.getWorkTime();
        int expected = 6;
        Assert.assertEquals(actual, expected);
    }

    @Test
    public void testCase2() {
        Mockito.when(weather.getWeather()).thenReturn("多云");
        int actual = work.getWorkTime();
        int expected = 7;
        Assert.assertEquals(actual, expected);
    }

    @Test
    public void testCase3() {
        Mockito.when(weather.getWeather()).thenReturn("阴");
        int actual = work.getWorkTime();
        int expected = 8;
        Assert.assertEquals(actual, expected);
    }

    @Test
    public void testCase4() {
        Mockito.when(weather.getWeather()).thenReturn("雨");
        int actual = work.getWorkTime();
        int expected = 5;
        Assert.assertEquals(actual, expected);
    }

}
```

由于依赖类的返回值是随机的，如果想通过有限的测试次数来测试 Work 类，则只能通过 Mock 进行测试。

下面对代码进行说明。

① @InjectMocks 注解用于创建被测类的对象，而@Mock 注解用于创建被 Mock 的对象。

② init()方法中调用了 Mockito 的初始化方法 initMocks()。

③ 通过 when().thenReturn()方式模拟 getWeather()方法，返回天气值。

④ getWorkTime()方法使用了模拟的天气值获取工作时间。

⑤ 使用 TestNG 的断言方法对预期和实际结果进行比较，完成断言。

修改 testng.xml 文件，在<test>标签中新增以下粗体部分内容：

```
<?xml version="1.0" encoding="UTF-8"?>
<!DOCTYPE suite SYSTEM "http://testng.org/testng-1.0.dtd">
<suite name="Suite">
    <test thread-count="5" name="Test">
        <classes>
            <class name="com.lujiatao.mocktest.WorkTest" />
        </classes>
    </test> <!-- Test -->
</suite> <!-- Suite -->
```

保存所做的修改，在 testng.xml 上用鼠标右击，从弹出的快捷单中选择“Run As → TestNG Suite”选项，然后查看测试报告，如图 11-1 所示。

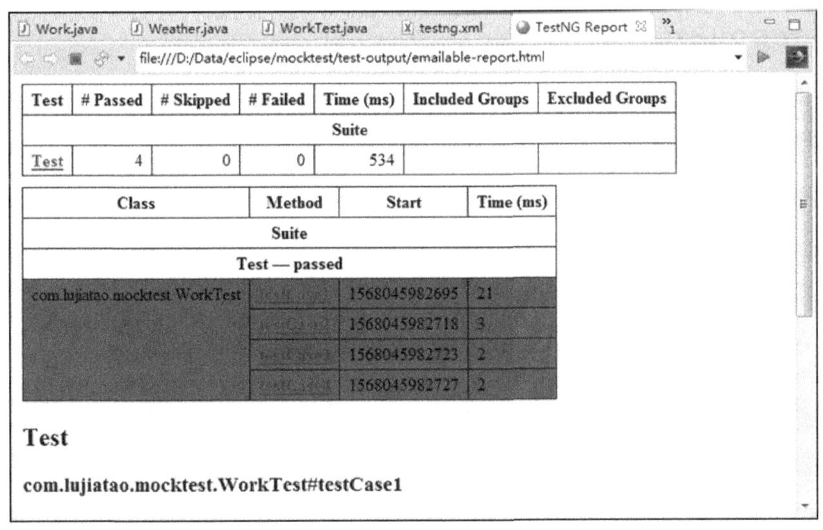

图 11-1

测试用例全部执行通过，与预期结果一致。

以上只演示了 Mockito 的简单用法，更多介绍参见官方文档。

11.2 接口 Mock 测试

11.2.1 接口 Mock 测试简介

接口 Mock 测试一般适用于以下两个场景。

场景一：在前后端分离的开发项目中，前端开发人员依赖后端开发人员提供的 HTTP 接口，此时后端 HTTP 接口尚未开发完成。

场景二：测试人员测试的系统依赖公司其他项目或第三方系统的 HTTP 接口。当公司其他项目的 HTTP 接口尚未开发完成时，需要进行接口 Mock 测试；当第三方系统的 HTTP 接口收费或者存在调用次数限制时，需要用接口 Mock 测试先将自身系统测试通过，再连接第三方系统做验收测试，从而达到节省开支或规避接口调用次数超限的目的。

接口 Mock 测试工具有很多，比如 Mock.js、MockServer 和 RAP2 等。RAP2 是 RAP 的升级版，由阿里巴巴的前端团队推出。有兴趣的读者可以访问它们的官方网站了解更多信息。

11.2.2 RAP2 用法

RAP2 可以使用在线版或本地版，在线版直接注册使用即可；本地版需要在公司内网搭建 RAP2 服务器，供公司内部使用。这里以在线版使用为例进行介绍。

1. 创建接口

访问 RAP2 官方网站可自动重定向到登录页面。

注册一个账号即可开始使用。在首页单击“新建仓库”，在“新建仓库”对话框中填写名称后即可完成仓库的新建（这里填写的名称为 Mock Test）。

进入新建的仓库，单击“新建接口”选项，在“新建接口”对话框中填写名称、地址、类型和状态码，完成接口的新建，如图 11-2 所示。

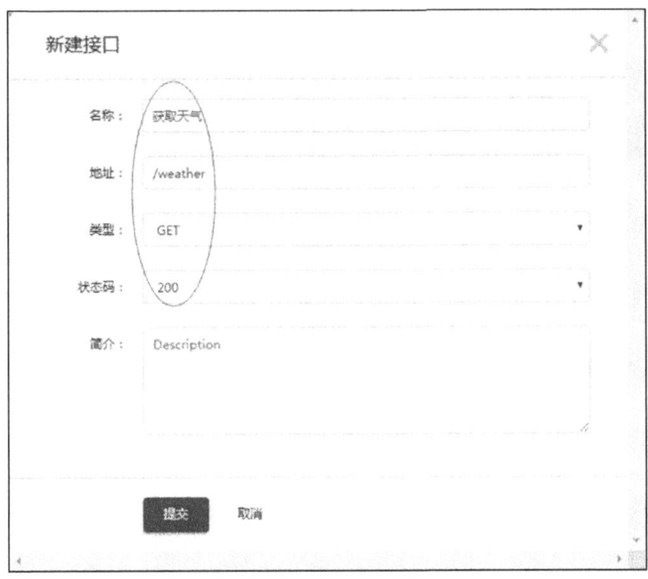

图 11-2

此时接口并没有响应内容，单击“编辑”选项，切换到“Body Params”页，配置响应内容后完成接口的配置，如图 11-3 所示。

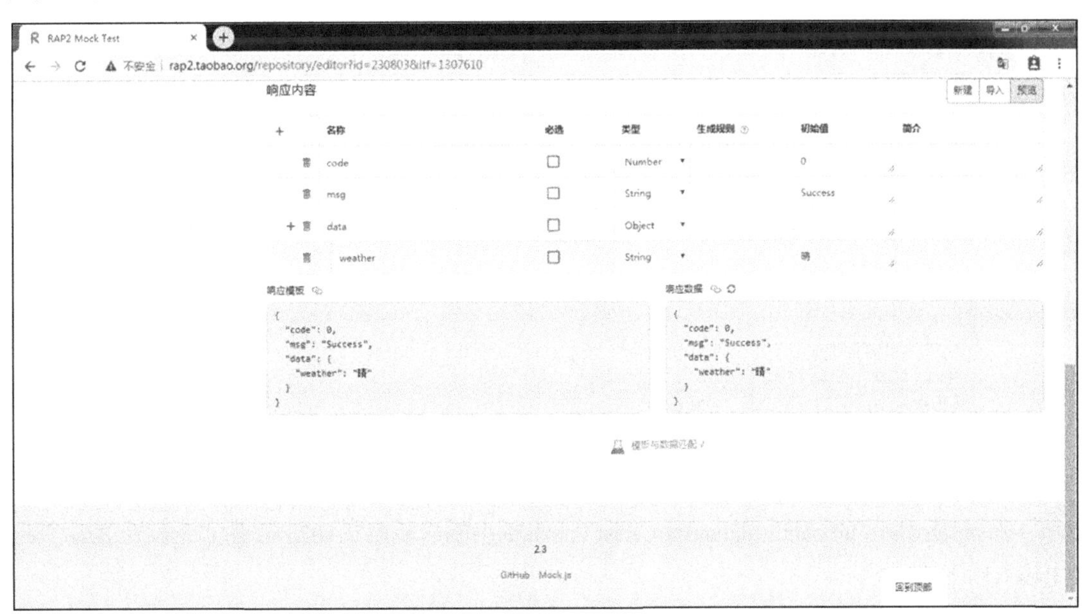

图 11-3

单击接口地址即可查看接口的响应内容，如图 11-4 和图 11-5 所示。

图 11-4

RAP2 Mock Test

rap2api.taobao.org/app/mock

不安全 | rap2api.taobao.org/app/mock/230803/weather


```
{
  "code": 0,
  "msg": "Success",
  "data": {
    "weather": "晴"
  }
}
```


图 11-5

2. 使用接口

修改 pom.xml 文件，在<dependencies>标签中新增以下内容：

```
<dependency>
    <groupId>org.apache.httpcomponents</groupId>
    <artifactId>httpclient</artifactId>
    <version>4.5.9</version>
</dependency>
<dependency>
    <groupId>org.json</groupId>
    <artifactId>json</artifactId>
    <version>20180813</version>
</dependency>
```

保存 pom.xml 文件，这时 Maven 会自动下载 HttpClient、JSON 及它们依赖的其他 jar 包。

在 src/test/java 的 com.lujiatao.mocktest Package 中创建名为 RAP2Test 的 Class，在 RAP2Test 中输入以下代码：

```
package com.lujiatao.mocktest;

import java.net.URI;

import org.apache.http.client.methods.CloseableHttpResponse;
```

```
import org.apache.http.client.methods.HttpGet;
import org.apache.http.client.utils.URIBuilder;
import org.apache.http.impl.client.CloseableHttpClient;
import org.apache.http.impl.client.HttpClients;
import org.apache.http.util.EntityUtils;
import org.json.JSONObject;
import org.testng.Assert;
import org.testng.annotations.AfterClass;
import org.testng.annotations.BeforeClass;
import org.testng.annotations.Test;

public class RAP2Test {

    private CloseableHttpClient client;
    private CloseableHttpResponse response;

    @BeforeClass
    public void init() {
        client = HttpClients.createDefault();
    }

    @Test
    public void testCase1() {
        JSONObject expectedData = new JSONObject().put("weather", "晴");
        JSONObject expected = new JSONObject().put("code", 0).put("msg",
            "Success").put("data", expectedData);
        JSONObject actual = null;
        try {
            URI uri = new URIBuilder().setScheme("http").
                setHost("rap2api.taobao.org")
                    .setPath("/app/mock/230803/weather").build();
            response = client.execute(new HttpGet(uri));
            actual = new JSONObject(EntityUtils.toString(response.getEntity()));
        } catch (Exception e) {
            e.printStackTrace();
        }
        Assert.assertEquals(expected.toString(), actual.toString());
    }

    @AfterClass
    public void clear() {
        try {
            response.close();
            client.close();
        } catch (Exception e) {
            e.printStackTrace();
        }
    }

}
```

这里使用了 HttpClient 来发送请求和处理响应，并且使用了第三方 JSON 库来处理数据，相关用法参见第 4 章。

修改 testng.xml 文件，修改<test>标签中的类，如以下粗体部分内容所示。

```
<?xml version="1.0" encoding="UTF-8"?>
<!DOCTYPE suite SYSTEM "http://testng.org/testng-1.0.dtd">
<suite name="Suite">
    <test thread-count="5" name="Test">
        <classes>
            <class name="com.lujiatao.mocktest.RAP2Test" />
        </classes>
    </test> <!-- Test -->
</suite> <!-- Suite -->
```

保存所做的修改，在 testng.xml 上用鼠标右击，在弹出的快捷菜单中选择“Run As → TestNG Suite” 选项，查看测试报告，如图 11-6 所示。

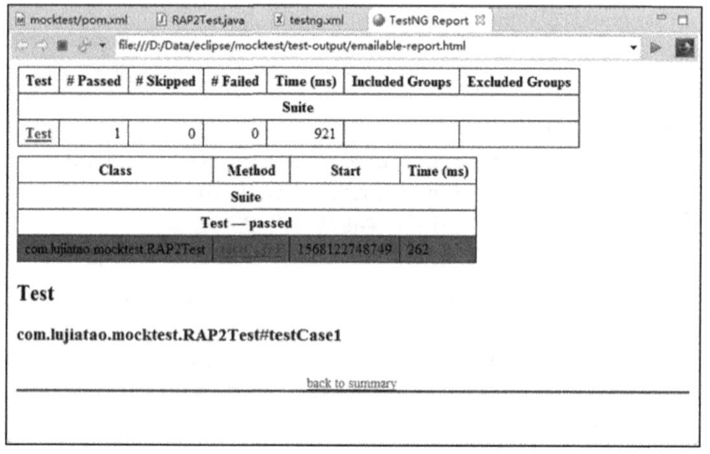

图 11-6

以上演示了一个很简单的接口模拟场景，在实际项目中，接口的复杂度往往更高，RAP2 提供了多种 Header、Query Param 和 Body Param 生成规则，生成规则的语法与 Mock.js 一致，具体可访问 GitHub，搜索 Mock，到 Mock.js 项目中查看语法规则。

11.3 代码覆盖率简介

代码覆盖率是软件测试的一种度量，是被测试覆盖到的源代码占总源代码的比例。

在 1.2.1 节中介绍了 6 种覆盖方法，即语句覆盖、分支（判定）覆盖、条件覆盖、分支（判定）—条件覆盖、条件组合覆盖和路径覆盖，通过这些覆盖方法算出来的覆盖率可作为代码覆盖率的统计指标。本章重点讲解代码覆盖率工具的使用，也会涉及其他代码覆盖率统计指标。常用的 Java 代码覆盖率工具如下。

Jtest：Parasoft 公司开发的商业测试工具，除可以进行单元测试外，还具有代码静态分析和代码覆盖率分析等功能。

JMockit：Mock 测试工具，可通过“埋点”的方式实现代码覆盖率分析，“埋点”方式对于待测系统而言是侵入式的。

JaCoCo：JaCoCo 是目前使用最广泛的 Java 代码覆盖率库之一，著名的 Eclipse 插件 EclEmma 便集成了 JaCoCo 库实现代码覆盖率分析。

有兴趣的读者可以访问它们的官方网站了解更多信息。

11.4　JaCoCo 用法

11.4.1　JaCoCo 计数器

在 JaCoCo 中，有自己的代码覆盖率统计指标，这些指标被称为计数器。

（1）Instructions（C0 Coverage）

这是最基本的计数器，统计单个 Java 字节码中指令的执行情况。

（2）Branches（C1 Coverage）

对 if 和 switch 分支进行统计，未覆盖被标记为红色菱形，部分覆盖被标记为黄色菱形，全覆盖被标记为绿色菱形。

（3）Cyclomatic Complexity

Cyclomatic Complexity 又称为圈复杂度，是软件模块结构复杂程度的一种度量方式。它是线性无关路径的数量，因此是应该测试的最小路径数。常用的基础路径测试法（Basis Path Testing，又称结构化测试）就是基于圈复杂度的一种测试方法。

圈复杂度 $v(G)$的公式如下：

$$v(\mathrm{G}) = E - N + 2$$

其中 E 是边数，N 是节点数。但 JaCoCo 使用了分支数（B）和决策点数（D）来代替，也就是说，使用以下等效方程可计算圈复杂度。

$$v(G) = B - D + 1$$

需要注意的是，JaCoCo 不将异常处理视为分支，因此 try-catch 语句不会增加代码的复杂度。

（4）Lines

未覆盖被标记为红色背景，部分覆盖被标记为黄色背景（代表只执行了该行的一部分指令），全覆盖被标记为绿色背景。

（5）Methods

该方法只要有一个指令被执行，就认为该方法被执行。

（6）Classes

该类只要有一个方法被执行，就认为该类被执行。

11.4.2 使用 EclEmma 插件

笔者的 Eclipse 已经默认集成了 EclEmma 插件，如果 Eclipse 中没有该插件，则按照安装 TestNG 插件的方法安装即可，此处不再赘述。需要说明的是，Work with 中输入的 URL 为 http://update.eclemma.org/，而不是 TestNG 的。安装完成后，在 Eclipse 工具栏可以看到如图 11-7 所示的图标。

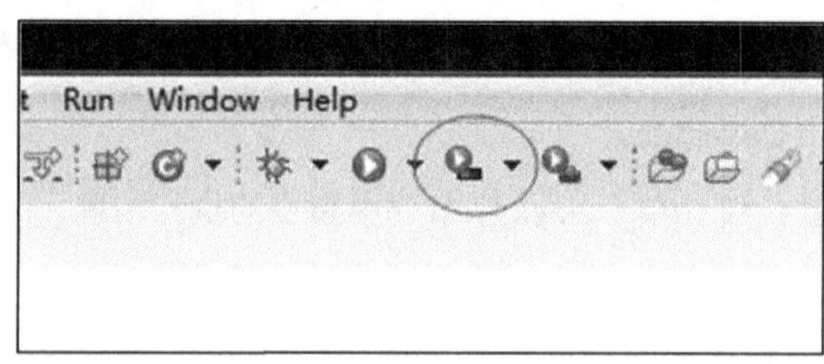

图 11-7

打开 Eclipse，创建一个新的 Maven 项目，关键信息填写如下。

Group Id：com.lujiatao

Artifact Id：codecoverage

Name：Code Coverage

当然，读者可根据实际情况填写，不需要和笔者填写的完全一致。

创建完成后，在 pom.xml 文件的<name>标签后输入以下粗体部分内容。

```
<project xmlns="http://maven.apache……"
    xmlns:xsi="http://www.w3……"
    xsi:schemaLocation="http://maven.apache…… http://maven.apache……">
    <modelVersion>4.0.0</modelVersion>

    <groupId>com.lujiatao</groupId>
    <artifactId>codecoverage</artifactId>
```

```
    <version>0.0.1-SNAPSHOT</version>
    <name>Code Coverage</name>

    <dependencies>
        <dependency>
            <groupId>org.testng</groupId>
            <artifactId>testng</artifactId>
            <version>6.14.3</version>
            <scope>test</scope>
        </dependency>
    </dependencies>

</project>
```

保存 pom.xml 文件，这时 Maven 会自动下载 TestNG 及其依赖的其他 jar 包。

首先创建被测类。在 src/main/java 中创建名为 com.lujiatao.codecoverage 的 Package 及名为 Work 的 Class，在 Work 中输入以下代码。

```
package com.lujiatao.codecoverage;

public class Work {

    public int getWorkTime(String weather) {
        switch (weather) {
        case "晴":
            return 6;
        case "多云":
            return 7;
        case "阴":
            return 8;
        default:
            return 5;
        }
    }

}
```

在被测类中有一个 getWorkTime()方法，该方法可根据不同的天气（入参）决定不同的工作时长（假设工作时长的单位为小时）。

然后创建测试类。在 src/test/java 中创建名为 com.lujiatao.codecoverage 的 Package 及名为 WorkTest 的 Class，在 WorkTest 中输入以下代码。

```
package com.lujiatao.codecoverage;

import org.testng.Assert;
import org.testng.annotations.BeforeClass;
import org.testng.annotations.Test;

public class WorkTest {
```

```
    private Work work;

    @BeforeClass
    public void init() {
        work = new Work();
    }

    @Test
    public void testCase1() {
        int actual = work.getWorkTime("晴");
        int expected = 6;
        Assert.assertEquals(actual, expected);
    }

    @Test
    public void testCase2() {
        int actual = work.getWorkTime("多云");
        int expected = 7;
        Assert.assertEquals(actual, expected);
    }

    @Test
    public void testCase3() {
        int actual = work.getWorkTime("阴");
        int expected = 8;
        Assert.assertEquals(actual, expected);
    }

    @Test
    public void testCase4() {
        int actual = work.getWorkTime("雨");
        int expected = 5;
        Assert.assertEquals(actual, expected);
    }

}
```

保存所做的修改，在 WorkTest.java 上用鼠标右击，从弹出的快捷菜单中选择“Coverage As → TestNG Test”选项，查看测试结果，如图 11-8 所示。

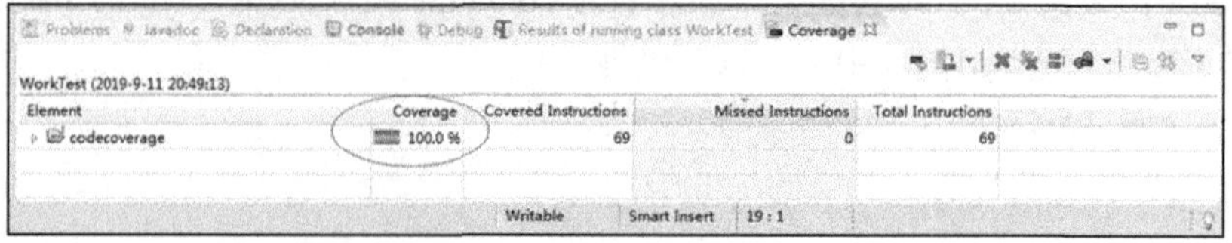

图 11-8

从测试结果可以看出覆盖率为 100%。

接下来删除 WorkTest.java 文件中的 testCase4，再次执行，此时覆盖率变成了 96.6%，如图 11-9 所示。

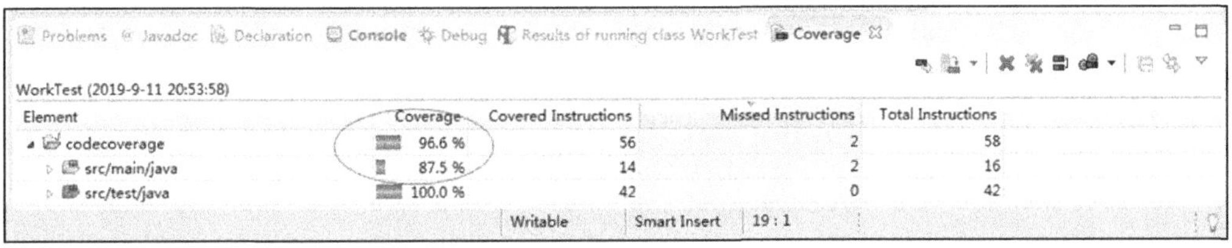

图 11-9

96.6%的覆盖率是如何计算出来的呢？将 Coverage 窗口放大，将测试结果展开如图 11-10 所示。

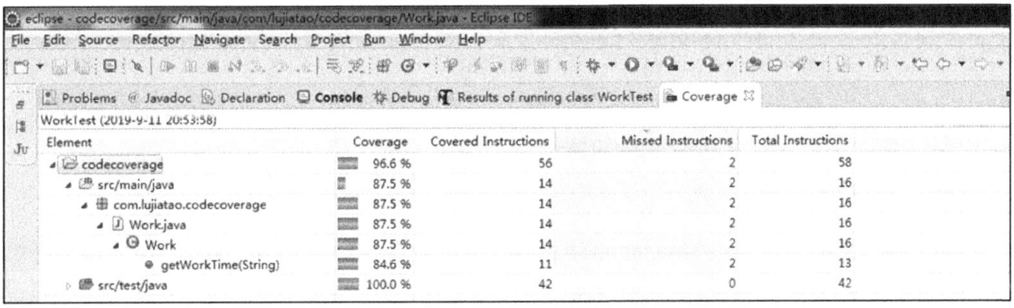

图 11-10

Covered Instructions 为 56，Missed Instructions 为 2（均来自 getWorkTime()方法），Total Instructions 为 58。计算方法如下：

```
Instructions (C0 Coverage) = Covered Instructions / Total Instructions × 100%
```

通过上面的公式可以计算出 Instructions（C0 Coverage）为 96.6%。

11.4.3 Maven 集成 JaCoCo

修改 pom.xml 文件，在<project>标签中加入以下内容。

```
<build>
    <plugins>
        <plugin>
            <groupId>org.apache.maven.plugins</groupId>
            <artifactId>maven-surefire-plugin</artifactId>
            <version>2.22.2</version>
            <configuration>
                <suiteXmlFiles>
                    <suiteXmlFile>testng.xml</suiteXmlFile>
                </suiteXmlFiles>
            </configuration>
```

```
            </plugin>
            <plugin>
                <groupId>org.jacoco</groupId>
                <artifactId>jacoco-maven-plugin</artifactId>
                <version>0.8.4</version>
                <executions>
                    <execution>
                        <goals>
                            <goal>prepare-agent</goal>
                        </goals>
                    </execution>
                    <execution>
                        <id>report</id>
                        <phase>test</phase>
                        <goals>
                            <goal>report</goal>
                        </goals>
                    </execution>
                </executions>
            </plugin>
        </plugins>
    </build>
```

保存 pom.xml 文件。

① maven-surefire-plugin：Maven 的测试插件，其中，在<suiteXmlFile>标签中指定了测试用例的 XML 配置文件。

② jacoco-maven-plugin：JaCoCo 的 Maven 插件，prepare-agent 代表使用 JaCoCo 的代理模式，另外，还定义了 JaCoCo 的测试报告。

在工程（codecoverage）上用鼠标右击，在弹出的快捷菜单中选择“TestNG → Convert to TestNG” 选项，在工程中生成 testng.xml 文件。

修改 testng.xml 文件，在<test>标签中新增以下粗体部分内容。

```
<?xml version="1.0" encoding="UTF-8"?>
<!DOCTYPE suite SYSTEM "http://testng.org/testng-1.0.dtd">
<suite name="Suite">
    <test thread-count="5" name="Test">
        <classes>
            <class name="com.lujiatao.codecoverage.WorkTest" />
        </classes>
    </test> <!-- Test -->
</suite> <!-- Suite -->
```

保存所做的修改，在工程（codecoverage）上用鼠标右击，在右键快捷菜单中选择“Run As → Maven test”，此时会先下载插件再进行测试。测试完成后进入 target/site/jacoco/目录，在 index.html 上用鼠标右击，在弹出的快捷菜单中选择“Open With → Web Browser” 选项 ，

JaCoCo 的测试报告如图 11-11 所示。

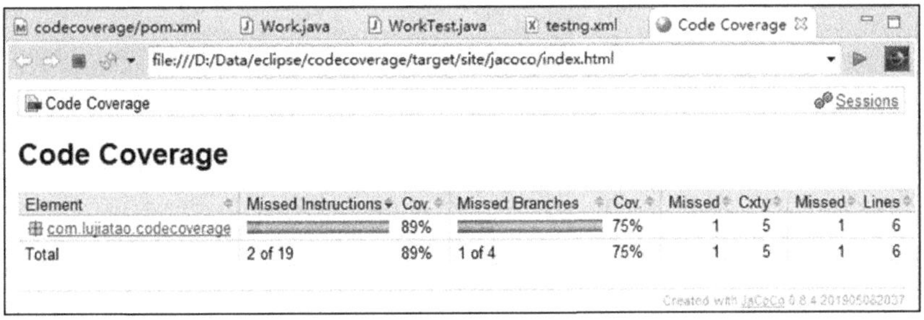

codecoverage/pom.xml　Work.java　WorkTest.java　testng.xml　Code Coverage

file:///D:/Data/eclipse/codecoverage/target/site/jacoco/index.html

Code Coverage　Sessions

Code Coverage

Element	Missed Instructions	Cov.	Missed Branches	Cov.	Missed	Cxty	Missed	Lines
com.lujiatao.codecoverage		89%		75%	1	5	1	6
Total	2 of 19	89%	1 of 4	75%	1	5	1	6

Created with JaCoCo 0.8.4.201905082037

图 11-11

这份测试报告显示了 JaCoCo 的 6 种计数器的结果，对于覆盖率不高的可以单击进去查看详情，最后可以定位到代码行，如图 11-12 所示。

codecoverage/pom.xml　Work.java　WorkTest.java　testng.xml　Work.java

file:///D:/Data/eclipse/codecoverage/target/site/jacoco/com.lujiatao.codecoverage/Work.java.h

Code Coverage > com.lujiatao.codecoverage > Work.java　Sessions

Work.java

```
package com.lujiatao.codecoverage;

public class Work {

    public int getWorkTime(String weather) {
        switch (weather) {
        case "□□":
            return 6;
        case "□□□□":
            return 7;
        case "□□":
            return 8;
        default:
            return 5;
        }
    }

}
```

Created with JaCoCo 0.8.4.201905082037

图 11-12

需要注意的是，在 Maven + JaCoCo + TestNG 这种组合中，如果 TestNG 中包含了断言且断言失败，那么会导致 Maven 构建失败（BUILD FAILURE），从而导致无法生成 JaCoCo 测试报告。

修改 WorkTest.java 文件中的 testCase3，将“expected = 8”改成“expected = 9”，再次执行 Maven test，控制台会打印错误信息，如图 11-13 所示。

```
[INFO]
[INFO] Results:
[INFO]
[ERROR] Failures:
[ERROR]   WorkTest.testCase3:34 expected [9] but found [8]
[INFO]
[ERROR] Tests run: 3, Failures: 1, Errors: 0, Skipped: 0
[INFO]
[INFO] ------------------------------------------------------------------------
[INFO] BUILD FAILURE
[INFO] ------------------------------------------------------------------------
[INFO] Total time:  5.863 s
[INFO] Finished at: 2019-09-11T22:23:14+08:00
[INFO] ------------------------------------------------------------------------
[ERROR] Failed to execute goal org.apache.maven.plugins:maven-surefire-plugin:2.22.2:test (default-test) on project codecoverage: There are test failures.
[ERROR]
[ERROR] Please refer to D:\Data\eclipse\codecoverage\target\surefire-reports for the individual test results.
[ERROR] Please refer to dump files (if any exist) [date].dump, [date]-jvmRun[N].dump and [date].dumpstream.
[ERROR] -> [Help 1]
[ERROR]
[ERROR] To see the full stack trace of the errors, re-run Maven with the -e switch.
[ERROR] Re-run Maven using the -X switch to enable full debug logging.
[ERROR]
```

图 11-13

当然，如果需要在测试失败时仍然生成 JaCoCo 测试报告，则可以通过修改 Maven 参数实现。

方法一：在 maven-surefire-plugin 的<configuration>标签中加入以下内容。

```
<testFailureIgnore>true</testFailureIgnore>
```

方法二：在执行 Maven test 时，在 Maven 命令中加入-Dmaven.test.failure.ignore=true 参数。

这样处理后，即使有测试用例执行失败，也不会影响 JaCoCo 测试报告的生成了，这一点从控制台打印的信息可以看出来，如图 11-14 所示。

```
[INFO]
[INFO] Results:
[INFO]
[ERROR] Failures:
[ERROR]   WorkTest.testCase3:34 expected [9] but found [8]
[INFO]
[ERROR] Tests run: 3, Failures: 1, Errors: 0, Skipped: 0
[INFO]
[ERROR] There are test failures.

Please refer to D:\Data\eclipse\codecoverage\target\surefire-reports for the individual test results.
Please refer to dump files (if any exist) [date].dump, [date]-jvmRun[N].dump and [date].dumpstream.
[INFO]
[INFO] --- jacoco-maven-plugin:0.8.4:report (report) @ codecoverage ---
[INFO] Loading execution data file D:\Data\eclipse\codecoverage\target\jacoco.exec
[INFO] Analyzed bundle 'Code Coverage' with 1 classes
[INFO] ------------------------------------------------------------------------
[INFO] BUILD SUCCESS
[INFO] ------------------------------------------------------------------------
[INFO] Total time:  6.113 s
[INFO] Finished at: 2019-09-11T22:50:00+08:00
[INFO] ------------------------------------------------------------------------
```

图 11-14